Design Added Value

Ömer Akın

Design Added Value

How Design Increases Value for Architects and Engineers

Springer

Ömer Akın
School of Architecture
Carnegie Mellon University
Pittsburgh, PA, USA

ISBN 978-3-030-28859-4 ISBN 978-3-030-28860-0 (eBook)
https://doi.org/10.1007/978-3-030-28860-0

This Springer imprint is published by the registered company Springer Nature Switzerland AG
The registered company address is: Gewerbestrasse 11, 6330 Cham, Switzerland

This book is dedicated to the thousands of students who have studied under Professor Ömer Akın over a span of 40 years. It was his life's work to support their endeavors and successes.

Foreword

Jim Garrett

Professor Ömer Akın and I were friends and colleagues for nearly three decades. When I returned to Carnegie Mellon University (CMU) in 1990 as a new faculty member in the Department of Civil Engineering, he quickly reached out to me as a fellow CMU alumnus (he from Architecture and me from Civil Engineering) to start a long and fruitful collaboration. The last time we met, in 2019 at a kebab restaurant in Squirrel Hill (near the CMU campus), we both greatly enjoyed the chance to catch up and discussed all the fun we had working together since our first meeting.

Ömer and I worked together with Professors Ulrich Flemming, Steve Fenves, and Rob Woodbury, and co-advised many doctoral students, on a project called the Software Environment to Support the Early Phases in Building Design (SEED). The SEED project sought to use object-oriented software concepts to create an architectural design support environment that supported the many phases from requirements elicitation and architectural programming (Ömer's focus) to building code evaluation (my focus). Following on the SEED project, we worked together, along with Professor Burcu Akinci and co-advised more doctoral students, on the creation of a software system to support the commissioning phase of building projects.

As further illustration of Ömer's CMU interdisciplinary DNA, he endeavored to create the Architecture Engineering and Construction Management program, a collaboration between the School of Architecture and the Department of Civil and Environmental Engineering, of which I was head at the time. The program continues to be a successful and valuable interdisciplinary graduate educational program.

It is these projects, and many others he worked on individually or with other colleagues, from which Ömer has gained the experience he relays in his last book "Design Added Value: How Design Increases Value for Architects and Engineers." Ömer provides an informed and passionate defense of the value that design brings to building projects, by describing the value-based design process and illustrating it with numerous effective examples.

One of my most favorite memories of Ömer was when he personally arranged for a small group of us, while attending a conference in Antalya, Turkey, to attend a performance of Verdi's opera Aida at an amazingly beautiful and acoustically functional Roman amphitheater in Aspendos. It spoke to what he most appreciated—effective design, beautiful architecture, his native home, and being with his family and friends. It also spoke so much about his love of life. Little did I know then how elegantly this experience would illustrate a key point of this book—effective design brings great value to any building project.

Carnegie Mellon University
Pittsburgh, PA, USA
January 2021

Jim Garrett

Foreword

Steve Lee

The earliest assignment of my academic career was teaching 1st year architecture studio with Professor Ömer Akın in 1981. This interaction had a lasting and profound impact on my attitude towards the studio and classroom. Over the years, Ömer and I went on to become good friends and through him I learned the importance of teaching *how* we design—not just *what* we design. Upon becoming head of the Carnegie Mellon University School of Architecture (CMU SoA) in 2008, I relied upon Ömer for guidance, support, and leadership in developing new programs for the school. Ömer had a positive and lasting impact on the SoA community—by my count there are over 2000 undergraduate and graduate students who truly benefitted from him as teacher, mentor, and/or agent provocateur.

It was with a heavy heart that day in 2020 that I learned of Ömer's passing after so many years of friendship. So, it is with great honor and respect for his many contributions that I write this foreword to, "Design Added Value: How Design Increases Value for Architects and Engineers."

Ömer was named Professor Emeritus in 2017, reflecting a faculty career at CMU that began in 1977. He earned his PhD in 1979 advised by Professors Charles Eastman, Bill Chase, and Herbert Simon and focused his research on design cognition, computer-aided design, and building commissioning. During his time with the school, he taught design studios and graduate courses, advised graduate students, and lectured both nationally and internationally.

Ömer developed the SoA's Architecture–Engineering–Construction Management (AECM) Master's and PhD degree programs jointly with the Department of Civil and Environmental Engineering, as well as the Doctor of Professional Practice (now the Doctor of Design) degree program. He served in many administrative positions, including Head of the Department of Architecture (now SoA) from 1981 to 1987.

A continuing thread through Ömer's teaching, research, and practice was the concept of value-based design (VBD)—the concept that design features have an intrinsic value that must be understood for every project. He utilized VBD in his practice with such world-renowned projects as the Turkish Nationality Room at the University of Pittsburgh, in his scholarly writings and in his teaching, and advising.

Ömer's required course, Ethics and Decision Making, was a fixture in the SoA for over three decades.

It is in the domain of design research then, that this new, posthumously published book belongs. It is written in a straightforward, declarative manner that I believe will be meaningful and accessible to architecture and engineering students. Part I is an introduction to the subject, Part II introduces numerical methods, Part III describes the methodology for design added value (DAV) analysis using two case studies, Part IV utilizes some well-known case studies to illustrate the concepts, and Part V brings it back to describing a designer's responsibilities.

As an educator, I appreciate the book's structure of narrative alternating with numerical methods and the fact that Ömer continuously refers to the role of the designer that makes this book truly engaging.

Thank you Ömer for your contributions as an educator, a researcher, and a practitioner to the world of design.

Respectfully, your friend and colleague,

School of Architecture, Carnegie Mellon University
Pittsburgh, PA, USA
January 2021

Stephen R. Lee

Preface

Design adds value!

Every time I've said this to a colleague they have responded with enthusiastic affirmation. This is the best-known fact in the world of design (including engineering, architecture, construction, urban, industrial, landscape, system, and software design) that does not have a body of research and literature attached to it. This is the nature of tacit knowledge. Yet, the more I looked at this phenomenon through notable examples the more convinced I became that the topic of added value of design deserves as much if not more attention than many other academic topics that occupy our attention.

The value design adds to a project is intrinsic value. Extrinsic value is the explicit cost of design assessed in the usual way. This is why assessors, patrons, and clients bulk at what they consider high budget figures that they consider excessive, disregarding the intrinsic value they could be adding to the value of their investments, only if they reconsider the costly design features initially proposed. Many good design ideas get ignored and thrown out during the budgeting process, or even worse, in the value engineering process, at the hand of assessors who fail to appreciate the intrinsic value of design features.

To wit, when Mr. E. J. Kauffman was having his *Fallingwater* built, he was concerned by the escalating costs. Today, Mr. Wright's masterpiece at Bear Run, PA, remains priceless.

While Mr. Utzon was effectively pushed out of his position as the architect of record of the Sydney Opera House over manufactured conflicts regarding the specialty plywood ordered to construct the acoustic panels, the remarkable edifice on the Sydney Harbor remains a symbol of the city if not the entire Commonwealth.[1]

Finally, who can assess the monetary value of the thrill of experiencing the canted top of the Citicorp Tower and its nine-story *pilotis* hovering above the Citicorp Plaza, in Manhattan, both features designed by William LeMessurier?

---

[1] Messent, David (1997) Opera House Act One David Messent Photography, Unit 17, 28 Rosenberry Street, Balgowlab, NSW 2093, Sydney Australia.

The thesis of this book is that intrinsic value of design features is indeed tangible value that needs to be explored, measured, and taken into account when initiating projects and financing their construction. It is not just some ephemeral and esoteric aspect of design, as some proponents of *Capital Project Delivery* (CPD) would have you believe. It is as calculable as the extrinsic value of a project. However, we need concepts, strategies, methods, techniques, and tools to do just that.

This text is dedicated to exposing and explaining these prerequisites of studying intrinsic value in buildings. We call this *Value-Based Design* (VBD) and the specific method of analysis that goes with it *Design Added Value* (DAV).[2]

To accomplish this goal we call not only upon both the sensibilities required by good taste but also, more importantly, upon all of the intellectual might we can muster for understanding enough about rational decision-making, Bayesian statistics, economics, cost–benefit analysis, data elicitation, pre and post facto analysis, expertise, creativity, planning, and optimization.

Enjoy!

Pittsburgh, PA — Ömer Akın
July 6, 2018

[2] Akın, Ö (2008) "Chapter 1: Current Trends and Future Direction in CAD" in CAD/GIS Integration: Existing and Emerging Solutions, edited by Hassan Karimi and Burcu Akinci, Taylor & Francis, NY, NY.

Acknowledgments

The work that led to the theory underlying this work has been supported by National Institute of Standards and Technology as well as the National Science Foundation.

The case studies and specific examples used have been developed by the author while teaching two required courses, *Value Based Design* and *Ethical Decision Making in Architecture,* in the School of Architecture and the Department of Civil Engineering, at Carnegie Mellon University. Students enrolled in these courses, during 2000–2016, were working towards their Bachelor of Architecture, Master of Science, and PhD degrees of the Architecture–Engineering–Construction Management (AECM) degree programs.

Two of the case studies have been developed by teams of students who have graciously released their copyrights and accordingly have been acknowledged in this text.

Special thanks to Professor Akın's former students Ipek Ozkaya and Yavuz Göncü for supporting this work to its fruition.

Contents

List of Figures

List of Tables

About the Author

Ömer Akın, PhD, AIA was Professor Emeritus of Architecture and Courtesy Faculty in the Department of Civil and Environmental Engineering at Carnegie Mellon University in Pittsburgh, PA.

Dr. Akın was also CEO of Architectural Design Associates (ADA), Inc. Dr. Akin published professional texts as well as fiction, including the following titles: Representation and Architecture (1982), Psychology of Architectural Design (1986, 1989), Generative CAD Systems (2005), A Cartesian Approach to Design Rationality (2006), Embedded Commissioning (2011), and Ethical Decision Making in Architecture (2018). His publishers include Information Dynamics Inc., Pion, Inc., Carnegie Mellon University Press, METU Press, Artech House, Inc., Springer, and CreateSpace, Inc.

Dr. Akın served as Professor of Architecture at Carnegie Mellon University, Pittsburgh, PA, USA, since 1978. He was a well-published researcher with several hundred reviewed publications, and texts that include the titles cited above among others. His research interests included design cognition, computer-aided design, case-based design instruction, ethical decision-making, value-based design, building commissioning, and automated design requirement management. He has also served as the Head of the School of Architecture and the director of the graduate programs.

Part I
Introduction to Design Added Value

This book aims to demonstrate that being smart about the economics of design can be the strongest vehicle for their designers to add value while enhancing their aesthetic qualities, disproving the old adage that you cannot have your cake once you have consumed it. The chapters collected under Part I will lay the foundational concepts that will guide design professionals in combining economics with the fields of architecture, engineering, and construction (AEC), establishing principles of value-based design.

Chapter 1
Introduction

Relevance of Economics

Economics is the study of assets and their value. It provides methods and tools to quantify the value of commodities, resources, and potentialities of goods available in the marketplace. Its relatively short history belies its rich and complex subject matter. The publication of Adam Smith's *The Wealth of Nations* in 1776 marks one of the tangible starting points of the field of Modern Economics. While Smith's seminal work refers to "Political Economy" and includes in its treatise the enrichment of both the sovereign and the people, after 1870, the term "economics" came into general usage, depicting a general-purpose approach to all assets and their transactions.

Smith, among other things, developed theories on the natural as opposed to the actual value of things. He is recognized for framing the notion of the *Invisible Hand* in opposition to that of division of labor that alludes to the forces and principles of conduct that transcend the total sum of individuals' rational decisions about economic transactions:

> *… how some overall pattern or design, which one would have thought had to be produced by an individual's or group's successful attempt to realize the pattern, instead was produced and maintained by a process that in no way had the overall pattern or design 'in mind.*[1]

Some of the earliest contributions in the development of modern economics also came through the works of Quesnay (1694–1774), Herman Conring (1606–1681), William Petty (1623–1687), Charles Davenant (1656–1714), and Godfried Achenwald (1719–1772) (Akin 2006). Demonstration of economic principles in more concrete terms motivated the emergence of new techniques for codifying quantifiable data. Statistics emerged as a tool from the works of Conring, who is credited with being the first to lecture on the subject. Petty and Davenant are credited with developing "Political Arithmetic." Achenwald used the word statistics for the first time, and Quesnay opened the door to using descriptive statistics to examine the "economics" of states (Arnold 1989).

[1] Smith A (1976) An Inquiry into the Nature and Causes of the Wealth of Nations (first published in 1776) R H Campbell and A S Skinner (general eds); W B Todd (ed), 2 volumes, Clarendon Press, Oxford

Ö. Akın, *Design Added Value*, https://doi.org/10.1007/978-3-030-28860-0_1

Advances in the methods used in economics originated from the development of concepts of "utility" and "value." These concepts were much debated during the "Scientific Socialism" movement of the nineteenth century, particularly in the works of Marx, Engels, Ranke, and Weber. This movement added significantly to the understanding of the norms of "society," impacts of Capitalism on society, and ultimately the "transformation of Hegel's philosophy of objective spirit into an analysis of human society." Subsequent work by Pareto on calibrating the value or utility of things and the economics of utility as agents of social structure led to the development of optimization as a means for "objectively" estimating the maximization or minimization of value (Arnold 1989).

Today, economics has branched out into every conceivable field of application and theory. In its own domain, it is broadly divided into micro- and macroeconomics. Microeconomics examines the economic behavior of agents (including individuals and firms) and their interactions through markets, subject to scarcity and government regulation. Macroeconomics examines behavior pertaining to national income and output, unemployment rate, price inflation, consumption, investment, spending, effects of monetary policy, fiscal policy, and their components. In its various diverse applications and investigations, the field of economics subsumes many definitions and distinctions which indicate alliances with a variety of other disciplines (Table 1.1). In this book, we investigate the interface of economics with several fields of design, bundled under the umbrella of architecture, engineering, and construction (AEC) of buildings and value-based design (VBD).

Interestingly, perhaps lamentably, social studies of practicing architects, particularly the most successful ones, care least about economics of their designs compared to other goals like aesthetics (MacKinnon 1970). Unfortunately, this attitude

Table 1.1 Diversity of subjects and distinctions within the field of Economics

Development economics	Financial economics	Marginalism
Classical economics	Game theory	Marxist economics
Cultural economics	Growth economics	mathematical economics
Behavioral economics and experimental economics	Heterodox economics: *history-institutions-social structure*	Mathematical: quantitative methods of economics
Econometrics	Schools of economics	National accounting
Economic geography	Industrial organization	Neoclassical economics
Economic history	Information economics	Normative economics
Economic reasoning	Institutional economics	Positive economics: *what is*
Economic systems	International economics	Prices and quantities
Economics in practice	Keynesian economics	Public finance
Economics' effect on society	Labor economics	Real-estate economics
Environmental economics	Law and economics	Supply-demand economics
Ethics and economics	Managerial economics	Welfare economics

is one of the reasons why architects are unable to make a stronger case for their aesthetics and design goals among other factors. This book aims to demonstrate that being smart about the economics of design can be the strongest vehicle for their designers to add value while enhancing their aesthetic qualities, disproving the old adage that you cannot have your cake once you have consumed it.

Architecture, Engineering, Construction (AEC)

Each a professional field engaged in creating buildings, in its own right, defines a broader but tightly knit field of practice often referred to through the acronym AEC (architecture, engineering, and construction). AEC represents a significant integration of multiple fields into one that is responsible for both the breadth and the depth of capital project design, delivery, and use. None of the components of this loosely knit domain of practice, by themselves, account for the entire life cycle of buildings, their infrastructure systems, and the many parallel but external practices that must be correlated to manage this entire life cycle.

The life cycle of buildings and their infrastructure systems have been characterized in a variety of ways by practitioners and researchers of the field. At a minimum, it includes a dozen distinct phases: requirement specification, preliminary design, design development, construction documents, construction, commissioning, occupancy, operations and maintenance, continuous commissioning, retrofitting, re-occupancy, and decommissioning. Furthermore, each of these phases requires input from a variety of design professionals and allied participants (Table 1.2). The professionals who need to conduct engineering tasks participate in design activities as design professionals in this life cycle.

Architecture and engineering fields account for the bulk of the design delivery tasks. They start with specifying the requirements of the facility and its programming. They go on to play a direct role in developing the design for the facility working through formalized phases and achieving collaboration between phases.

Their role goes well into the commissioning and occupancy of buildings if not their retrofitting and re-commissioning, over time. In most cases, contractors are also involved early on due to the growing consciousness about the improved management of the construction process. The early collaboration between contractors and design architects and engineers has proven to be invaluable in making the delivery of a buildings' budget and schedule more robust and reliable. Contractors' role is central also later, during the commissioning, operations, and maintenance stages. Finally, in the decommissioning stage, again, they play a key role.

Table 1.2 Participants in the delivery and life cycle use of buildings (Akın 2006)

Building life cycle	Design professionals	Other participants
Requirement specification	Architects, facility programmers, planners, appraisers, specialty designers	Clients, users, owners, regulatory agencies
Preliminary design	Architects, civil engineers, mechanical engineers, planners, cost analysts, specialty designers	Clients, users, owners
Design development	Architects, civil engineers, mechanical engineers, specialty designers	Clients, users, owners, regulatory agencies
Construction documents	Architects, civil engineers, mechanical engineers, data management specialists, contractors, specialty designers	Regulatory agencies
construction	Contractors, subcontractors, specialty manufacturers, architects, civil engineers, mechanical engineers, data specialists	Regulatory officials, local officials, owners
Commissioning	Commissioning authority, contractors, mechanical engineers, architects, operation and maintenance staff	Users, clients, owners
Occupancy	Contractors, space planners, interior designers	Users, clients, owners
Operations and maintenance	Contractors, operations and maintenance staff	Users, clients, owners
Continuous commissioning	Commissioning authority, mechanical engineers, architects, operation and maintenance staff	Users, clients, owners
Retrofitting	Architects, facility programmers, planners, contractors, appraisers	Clients, users, owners, regulatory agencies
Re-occupancy	Space planners, interior designers	Users, clients, owners
Decommissioning	Contractors, planners, appraisers, developers	Users, clients, owners, regulatory agencies

Value-Based Design (VBD): Combining Economics and AEC

As stated in the *Preface* section, the central focus of this book is the value that design adds to the building life cycle process in both quantitative and qualitative terms. The principal logic of VBD is that design with the increase of monitory or material resources for the project can be, and often is, the sole reason for the increase in the *Net Present Value* (NPV) of a capital project. The value of this cost differential,[2] which is desirable if it is positive and dreaded when it is negative, has been tacitly used in design decision-making since time immemorial.

Our aim is to make the knowledge about the definition, calculation, and application of VBD explicit and to develop reliable methods for estimation, prediction, and

[2] Value of cash inflows and the value of cash outflows over a period of time.

control of value added to buildings and other facilities through design. Surely, the field of economics is a principal ally in this task. Furthermore, economics is no stranger to the building and facility life cycle cost estimates.

Real-estate economics (Table 1.2), design economics, and *Building Economics* (Bon 1989; Wilson 1990) are some of the titles that have been used to frame this area of knowledge, with considerable impact. Some of the economics applications in the AEC fields that have become commonplace includes construction costing and budgeting, *value engineering*,[3] and investment planning and appraisal. The underlying premise of these methods and approaches is to make sure that the economic value aspects of AEC tasks are managed with an eye towards effective results of design quality, investment, and revenue.

While acknowledging their importance for the AEC fields, this book approaches the interface of AEC and economics with a premise quite different than value engineering and budget assessment practices. We claim that design adds value to buildings and facilities beyond those estimated by conventional methods, and often this value goes unnoticed until most if not all financial decisions are already made. This has several deleterious effects that need to be recognized.

Design value goes unnoticed when the economic impact of VBD is not explicitly estimated. In this case, if the VBD would turn out to be positive, the value adding feature would not have been realized. This is opportunity loss. In the event that VBD would turn out to be negative, the realized features would lead to design value gold plating and consequently unneeded added cost. In both cases, there is the potential of monetary loss compounded by loss of reputation and opportunity.

Currently there is no set methodology for carrying out VBD estimates. Often, the economic value of a novel design feature is inaccurately estimated. This potentially leads to the same two outcomes cited above: lost design opportunity value or design value gold plating.

In the remainder of this book, we will provide many examples of such cases, ranging from Fallingwater at one end of the spectrum to Kansas City Hyatt Regency on the other. We will cover these two cases in detail, the former one illustrating the gains and the latter illustrating the losses that can be measured through VBD methodology. Our purpose is to use the case studies in this text to substantiate the argument that VBD can provide benefits to all parties involved in the building life cycle process, especially to clients and owners of these facilities.

Chapters 1–4 of Part I, in combination, constitute an introduction to the VBD methodology than the analytical techniques covered in Chaps. 5–9 of Part II. These techniques are illustrated with detailed DAV analysis examples in Chaps. 10 and 11. The supplemental case studies of Chaps. 12–17 of Part IV provide an opportunity to apply the analytical skills gained from the earlier chapters through exercises provided at the end of each chapter.

[3] A concept quite different from value based design as we will explain later.

Chapter 2
Capital Projects and Building Assessment

Capital Projects

Arguably, the most significant personal expenditure we ever make, both in terms of absolute numbers and as ratios of total assets we hold, is building related (Tables 2.1 and 2.2). Housing makes up about 34% of the total expenditures, and this percentage has stayed consistently this high over the decades. The potential impact of capital projects on society and individuals is enormous. In addition, inefficient design, operations, and management of buildings have other tangible costs, such as wasteful energy use and harmful environmental consequences.

> *We no longer build buildings like we used to, nor do we pay for them in the same way. Buildings today are... life support systems, communication terminals, data manufacturing centers, and much more. They are incredibly expensive tools that must be constantly adjusted to function efficiently. The economics of building has become as complex as its design.*[1]

According to the Department of Labor, Bureau of Labor Statistics, published in 2017, the average annual expenditures of all consumer units in the USA is $60,060. Out of this, by far the largest component is for housing (30–38%), with transportation and food as distant second and third, respectively, at 17% and 12.8% (Fig. 2.1).

Outside of the housing sector, principally the industry, commerce, and transportation sectors, as well as a myriad of others to a lesser degree, also contribute significantly to the construction, operations, and maintenance of buildings and their physical infrastructure. One estimate places it upwards of 40% of the annual GDP. However we slice it, the importance of buildings in the economic life of the USA comes up paramount.

Conservative estimates of expenditure related to energy consumption in the AEC sector range between 25% and 36% of all energy expenditures in the USA. The

[1]Wilson R (1990) Foreword. In: Ruegg RT, Marshall HE (eds) Building economics: theory and practice. Van Nostrand Reinhold, New York, NY.

Ö. Akın, *Design Added Value*, https://doi.org/10.1007/978-3-030-28860-0_2

Table 2.1 Average annual expenditures, 2008 (DLBLS 2010)

Expenditures	Amount	Percentage
Food	$6,443	12.8
Alcoholic beverages	$444	0.9
Housing	$17,109	33.9
Apparel and services	$1,801	3.6
Transportation	$8,604	17.0
Healthcare	$2,976	5.9
Entertainment	$2,835	5.6
Personal care products and services	$616	1.2
Reading	$116	0.2
Education	$1,046	2.1
Tobacco products and smoking supplies	$317	0.6
Miscellaneous	$840	1.7
Cash contributions	$1,737	3.4
Personal insurance and pensions	$5,605	11.1
Total	$50,486	100

Table 2.2 Housing expenditure subcomponents (DLBLS 2010)

Subcomponent	Amount	Percentage
Shelter	$10,183	59.5
Utilities, fuels, and public services	$3649	21.3
Household operations	$998	5.8
Housekeeping supplies	$654	3.8
Household furnishings and equipment	$1624	9.5

lower bound corresponds to 300.87 billion dollars per year. This is no different in other developed economies around the globe, so much so that the measure of human development has become synonymous with the development of capital projects. This is the case whether considering the emergence of new nexus for development, like Dubai in the Middle East and Shanghai in China, or the description of civilizations, as in ancient Egypt, or Rome.

The construction of physical infrastructures and buildings is often the most prevalent testimonials to the advancement of society. By the same token, the economic investment in them is undeniably and unavoidably a serious commitment that needs clarity and understanding. Just including three categories: residential, industrial, and commercial, in which building professionals are the primary decision-makers of design and operations, the total expenditure prorated to the consumption in each category is over 300 billion dollars in the USA. This is roughly 25% of all expenditures in all sectors. This conservative estimate does not include the building activity in the transportation sector or myriad of relatively smaller sectors like office, education, healthcare, warehousing and storage, lodging, food service, food sales, public assembly, and the service sector. *Trends in Building-Related Energy and Carbon Emissions*

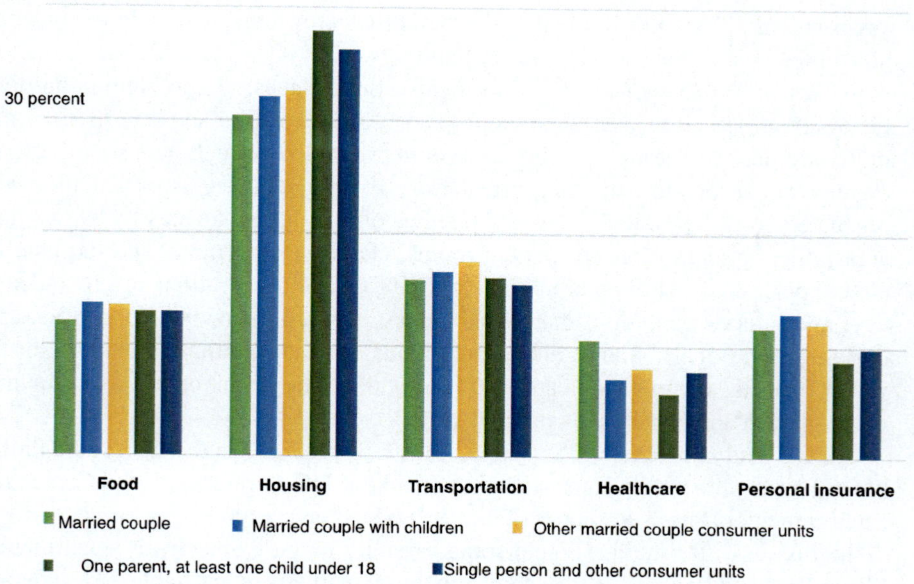

Fig. 2.1 Shares of average annual expenditures on selected major components by composition of consumer unit, 2017

estimates the energy consumption in the housing combined with the commercial sectors to be 36% of all US consumption (Battles and Burns 2000).

Evaluation in the AEC Sector

Given the criticalities introduced by inherent expenditures seen in the AEC sector by individuals, societies, and the environment, it appears that evaluating and monitoring capital project delivery should be a key function of this sector. Let us now briefly review some of the current methods of evaluation available in the AEC sector.

Post-Occupancy Evaluation (POE)

POE is a practice that started in the USA and primarily recognized with this acronym mostly in the USA and Canada. It is "*the process of evaluating buildings in a systematic and rigorous manner after they have been built and occupied for some time*" (Preiser 2018). It has a broad scope, which is usually customized to fit the project at hand. It can be specifically targeted to a performance issue like daylighting or energy consumption as well as broadly applied to the diagnosis of other potential problems seen in capital projects. Since its inception through the environmental design movement of the 60s, it has grown and evolved into a recognized life cycle evaluation method for buildings.

A typical, broad scoped POE will involve three stages: (1) problem definition, (2) observation and measurement, and (3) data interpretation and reporting. Some POEs are defined through performance dimensions like visual, acoustic, thermal, organizational, privacy, lighting, orientation, and maintenance aspects. Others are organized around physically situated features of a building, such as lobby, waiting areas, dinning and cafeterias, patient rooms, classrooms, terraces, and the like. In either approach, a set of methods can be used to make observations in situ and surveys and interviews with primary stakeholders: occupants, owners, operators, and design professionals. Some of this data lends itself to qualitative analysis as in visual aspects, comfort indicators, privacy, and interaction between occupants, while others can be analyzed quantitatively.

A POE study that exemplifies some of the best practices in the field is the Philip Merrill Environmental Center study conducted at UC Berkeley, by the Center for Environmental Design Research (Table 2.3) (Heerwagen 2005).

In this study, the authors found some variables that are objectively quantifiable, like airflow and light levels, and some others subjectively quantifiable, like satisfaction with air quality and light levels:

- Occupants were highly satisfied with the building as a whole.
- Occupants gave a positive response to air quality.
- 90% of occupants were satisfied with daylighting.
- With 80% of the occupants, all psychosocial ratings were positive.
- Occupants expressed a strong sense of pride with the building.
- Acoustical conditions were negatively rated, due to speech privacy.

Table 2.3 POE performance measures at the Philip Merrill Environmental Center—average scores by category ($N = 71$) on a seven-point scale of −3 through +3

General satisfaction—building	2.3
General satisfaction—Workspace	2.0
Office layout	1.3
Office furnishings	2.2
Thermal comfort	0.6
Air quality	2.1
Lighting	1.8
Views	1.7
Acoustic quality	1.0
Cleanliness and maintenance	1.5
Attention and concentration	1.0
Awareness and communication	1.1
Interactive behavior	1.3
Functionality	1.7
Acoustic functionality	0.3
Community	1.8
Morale and well-being	1.5

Facility Performance Evaluation

Recently, research institutions that teamed up with government agencies like General Services Administration redefined the POE process as *Facility Performance Evaluation* (FPE). The distinction is in the methodological clarity and rigor that is promoted by FPE. In place of a customized approach, an a priori set of tools and techniques are employed to discover faults and substandard performance aspects in buildings when following the FPE process (Zimring et al. 2010).

Building Feasibility Analysis

Building Feasibility Analysis (BFA) studies also bear a resemblance to POE and FPE. BFA are performed on commissioned facilities to specify the requirements for a replacement facility or to determine the feasibility of a new design. While its purpose is somewhat different from POE and FPE, its agenda is similar.

In a BFA, the factors to consider include "*given building facilities, physical layout of the space, building systems: HVAC, electrical, communication, structural and technical constrains, location, neighborhood, parking facilities, lease terms, demolition and improvement costs necessary to update building, client's budget constraints, building codes, and Americans for Disabilities Act (ADA) compliance requirements. If more than one building is considered, then the comparative analysis between the different sites becomes necessary.*"[2]

Simulation-Based Evaluation

The use of simulation as an evaluative technique is common in *Close-Fit Buildings*[3] (Akın et al. 2012) where advanced technology is used in designing the Mechanical, Electrical, Plumbing (MEP) systems. Without simulation-based techniques in such buildings, it becomes almost impossible to connect the dots between the plethora of settings generated by controls systems that are governed by proprietary software composed of subcomponents and drive air handling units, heat exchangers, and variable-air-volume boxes controlling the environmental conditioning in a building.

---

[2] Warszawski A, Rosenfeld Y (1994) Robot for interior-finishing works in building: feasibility analysis. J Constr Eng Manag 120:1.

[3] Close-fit buildings are buildings that use sophisticated systems and complex technologies to satisfy their missions. These buildings rely on well-informed users, abundant resources, and advanced operations and management (O&M) procedures to closely meet the performance requirements defined at the outset.

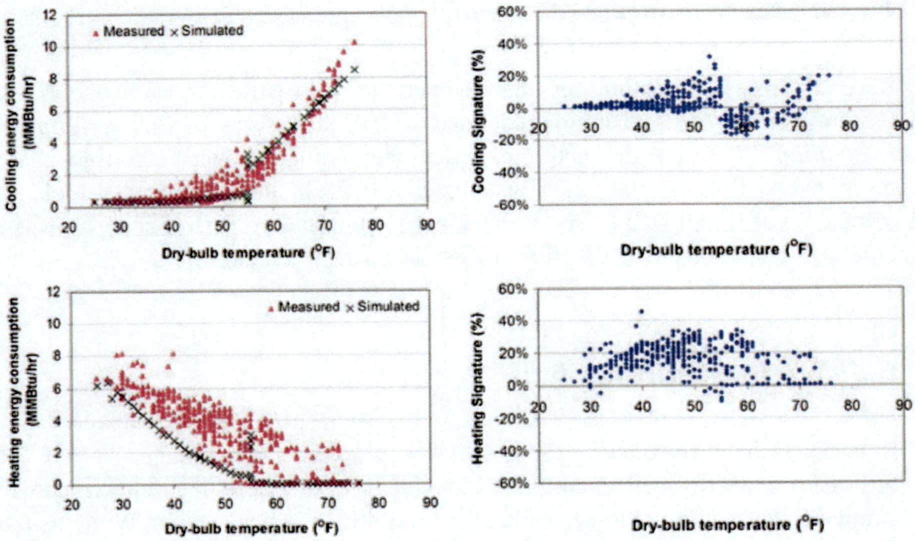

Fig. 2.2 "Comparison of measured cooling and heating consumption with initial simulated values and initial cooling and heating calibration signatures." (Liu and Liu 2011a, b)

Simulations provide the benchmarks against which the measurements obtained from the field can be validated and accurately interpreted (Fig. 2.2).

Other applications of building system simulation include self-configuring systems (Akin et al. 2009), egress patterns (Pan et al. 2006; Pan et al. 2006; Rahman et al. 2008), biological analogs (Westre 2008), and productivity modeling (Hong et al. 2000).

The rationale for computer simulations in building evaluation arises from the need to accurately estimate, predict, and control the parameters, such as pressure, temperature, energy, and fluid flow rate in functioning HVAC system, in their dynamic state. This requires that in addition to the attributes of the building system, its response to changing external conditions be accurately modeled. To accomplish this, the results of a steady-state simulation of the system's energy and control flow values are known. For more complex cases, including lighting, acoustic, and ventilation flows, finite element analysis simulations are needed.

The principal concept underlying digital simulation applications is to create sufficiently realistic virtual representations of phenomena to enable accurate calculations of how well they would perform in reality, that is, if they were constructed as designed. This approach has the enormous advantage of assessing performance prior to significant capital expenditures and the discovery of mistakes before they are realized. This is a growing area of practice that will remain an important aspect of building evaluation.

Building Commissioning

Arguably the gold standard of capital project evaluation is Building Commissioning (BCx) and Embedded Commissioning (ECx) (Akin et al. 2012). These techniques help us understand that the greatest cost and performance gains can be attained by improving energy-consuming subsystems in buildings and their infrastructures. BCx is a cost-effective way of determining the inefficient use of energy in HVAC and lighting systems of buildings, related to direct energy savings, greenhouse emissions, occupant comfort, indoor air quality,[4] and reduced operational costs. It is one of the most systematic, accurate, and advanced building evaluation approaches applied to the energy-consuming subsystems.

ECx is the more advanced digital form of BCx designed for the information age. "[It] *is an information-based technology applied longitudinally to the building life-cycle process. Its purpose is to embed every stage of the capital project delivery (CPD) process, from requirement specification to operation, with interoperable and persistent information"*[5] such that continuous commissioning of building systems becomes routine.

Current Challenges and Opportunities in the AEC Sector

Arguably, the AEC industry presents a dynamic and at times unpredicTable picture to the outside world. Cost, schedule, and design quality, the three indispensable pillars of project management, are often compromised beyond the level of acceptability that is standard in other sectors of global industries.

Peter Beck (2001) states: "*Over the past three decades, most industries have undergone significant transformations resulting in substantial improvements in the value of their products and services. Automobile manufacturers, for instance, have reduced their concept-to-production cycle from six years to 14 months. Wal-Mart has revolutionized retailing by developing a highly sophisticated business model enabled by a powerful distribution technology. Mini-mills have redefined the steel industry by deploying smaller, efficient operations, using new technology, and Amazon has made purchase of goods over the internet routine. In most cases, these extraordinary improvements resulted from significant changes in both business models and the processes they include, frequently enabled as a result of some technological innovation.*"[6]

[4]Indoor Air Quality Guide Best Practices for Design, Construction, and Commissioning; ©2009 American Society of Heating, Refrigerating and Air-Conditioning Engineers, Inc. 1791 Tullie Circle, NE Atlanta, GA 30329 www.ashraeorg

[5]Akin ., Akıncı B, Berges M, Bushby S, Garrett J, Huber D, Lee SH, Türkaslan-Bülbül T (2012) Embedded commissioning of building systems. Artech House, Inc., Boston, MA

[6]Beck, P., "The AEC Dilemma: Exploring the Barriers to Change," Proc. Design Intelligence, Greenway Communications and Design Futures Council, Washington, DC, February I, 2001, http://www.di.net/articles/archive/2046/

The Capital Projects Industry on the other hand is recognized for its inability to realize such improvements. Teicholtz writes "*A building that took 1,000 hours to construct in 1964 would have required just 552 hours in 1998 had the industry achieved the same productivity increases as the rest of the non-farm sector. Instead, that building took more than twice as many hours: 1,185.*"[7] It is predicted that, in the US construction market, schedule and cost overruns are in the order of $200–500 billion (LePatner 2008).

Several players in the construction sector have taken this challenge on with new management strategies and digital tools for the improvement of CPD performance. The *Lean Construction* approach investigates the advantages of engineered-to-order products, cost accounting decision-making, integrated products and processes, designing for target cost, work flow reliability, and managing materials at the construction site (Ballard 2008). Construction Industry Institute sponsors research on similar topics, focusing on a broader agenda of construction performance issues documenting the wasteful practices of CPD (CII 2009).

Fieldworker's productivity losses in the building sector have been documented through shadowing studies in an institutional context, to be anywhere between 5 and 20% (Lee and Akın 2010). According to Beck (2001), current AEC practices may be responsible for 25% in-field inefficiencies due to incomplete and poorly coordinated contract documents, excessive reliance on shop drawings and *Request For Information* (RFI) procedures, change orders, outdated manufacturer specifications, value engineering to reduce scope, and project managers devoting about 60% of their time to checking, fixing, and documenting problems (Beck 2001).

Micro-level studies on change order costs in institutional construction projects place the added cost at 8–10% of total cost (Akın and Anadol 1993). Prorated to 2010 dollars, this is equivalent to $108 billion—calculated as "total construction sector expenditures in 1998," which makes it the largest component (40%) of project cost overruns. In addressing issues like this in the AEC industry, we recognize several key concepts that will shape the emerging new practices including performance emphasis, stakeholder savvy, digital information use, and integration of design and operations.[8]

Performance Emphasis

In the interest of improving accuracy and quality in CPD there has been a shift in requirement specification and evaluation techniques. Prescriptive building specifications are being abandoned, or at least augmented, in lieu of performance-based

[7] Teicholz, Paul (2004) "Labor Productivity Declines in the Construction Industry: Causes and Remedies," AECbytes Analysis, Research, Reviews, Web Publication, Issue #4 (April 14) http://www.aecbytes.com/viewpoints.html

[8] Value Based Design, a primer for the Value Based Design Course (48–759) ~ School of Architecture, Carnegie Mellon University, Pittsburgh, PA (2005–2016).

specifications especially in areas like energy-consuming systems (Debajyoti et al. 2006), construction methods (Lemay 2004), and construction safety (USDT-FHA 2004), among others.

Building specifications and requirements, when performance-based, emphasize functional metrics for design and construction. Means of construction and ends of design are measured through process and functional description. Solutions and processes are not prescribed. Designers and contractors accomplish desirable performance outcomes in whatever way they deem best. While descriptive specifications can lead to innovation and increased quality, it can also mean that the bottom line can drag designs and buildings to the lowest possible state allowable by the performance range. Prescribed solutions, on the other hand, can be overdesigned, and, by the same token, avoid the least accepTable outcomes.

Stakeholder Savvy

Stakeholders are the owners, users, financiers, and developers of buildings and infrastructure systems. Increasingly, stakeholders assume greater responsibility in all phases of building delivery. In the programming phase when the requirements are specified, they focus on meeting certain performance metrics, which include conservation of energy, carbon emissions, and remediation of indoor air quality. Increasingly, stakeholders also consider improving occupant performance levels.

After the commissioning of buildings and building infrastructures, owners play an increasingly active role. Institutional clients usually have their own facilities departments and closely monitor the performance indicators of their capital projects. Common critical performance issues consist of achieving energy efficiency and reducing carbon footprint. However, there are special performance considerations in facilities with critical functions such as laboratories, industrial plants, prisons, hospitals, and transportation hubs with special performance requirements like security, uninterrupted operations, egress control, and flexibility of operations that are considered secondary. As a result of heightened awareness of specialized requirements, capital project performance measurement and evaluation have become more frequent, sophisticated, and accessible. Clients and users require, even demand, environments that match and exceed their requirements of high productivity, low cost, environmental sensitivity, and agility in use. They expect buildings and infrastructures to add value through their methods of design and construction, which influence the extend buildings and infrastructures can satisfy their general and specialized performance requirements.

Integration of Design and Operations of Capital Projects

A key benefit of the digital realm is to reduce costs and delays stemming from the iterative and repeated creation, and recreation, of information sets specific to ever narrowing phases of the CPD. Manual handoff and disconnected information acquisition practices create spikes in information loss and gathering at the points of transition from one phase to the next (Fig. 2.3).

The saw-tooth-shaped curve in Fig. 2.3 indicates the expenditure of resources in the manual mode, over time. Each sharp increase marks a shift from one phase of the CPD process to the subsequent one. The resources used to develop information are often in formats suiTable only to the earlier phase and does not support work in the subsequent phases. In each new phase, there is a sharp increase in the number of resources committed. The smoother curve depicts the resources required by digital tools and practices in the same CPD phase.

When using digital tools, data representation is interoperable, that is transferable with ease from one phase to the next. The commitment of resources is less than the manual mode owing to the reuse of data that has been generated in earlier phases. These two modes of practice represent the several trade-offs of CPD. Either pay in smaller amounts but frequently throughout the CPD process or pay a lot, up front while paying less overall, due to the diminishing need for additional resources, in the later phases.[9]

In building commissioning, it is documented that 33% of all faults detected deal with construction and installation, while design accounts for 35%. In other words, a total of 68% of faults detected in the operations phase could have been

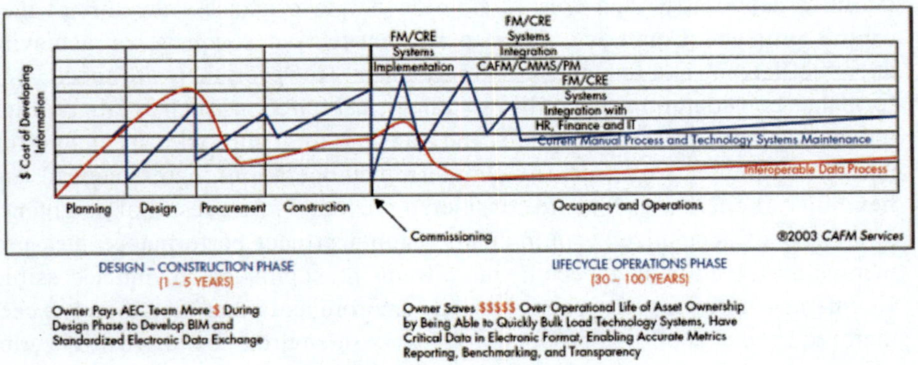

Fig. 2.3 Saw-tooth model of information acquisition and loss in CPD—"Pay now or pay later" diagram motivating interoperable and persistent information models for the Architecture-Engineering-Construction (AEC) industry. Courtesy of Andy Fuhrman, International Facility Management Association, 1 E. Greenway Plaza, Suite 1100, Houston, TX

[9] http://www.iea-ebcorg

prevented in an earlier phase. Robust information that connects all of the phases can help diagnose and properly remedy faults through feedback loops. The digital information base of a project that is truly interoperable can provide feedback as well as feed-forward of information which is just as crucial. A digital information base and feed-forward information ensures efficiency in data acquisition and reduces loss of data that is already acquired and is instrumental in fixing faults at the source whether it is in the requirement, design, or construction phase of the CPD.

BIM: The Emerging Digital Platform of Capital Project Delivery

At the heels of the growing interest in modeling and visualization of complex building projects of the 1990s, BIM (building information modeling) came into being (Eastman et al. 2008). Assisted by the Object Oriented (OO) modeling paradigms of software engineering and the use of advanced computer hardware, BIM aspires to organize "*all of the information about buildings in a multi-dimensional space, … containing information on topics as diverse as lifecycle of building delivery, specialized building systems consultants, product and process coordination, cost-schedule-quality measures, … subsuming a space that is governed by shared product and process representations conforming to broadly accepted standards*".[10]

BIM is shaped by three major influences: large data management, intra-task collaboration, and smart representations. Capital projects have become very large, the most modest versions involving hundreds of thousands of modeled objects. Ease in modeling and communication with other professionals without information loss of such a large repertoire of objects gives impetus to BIM applications, where syntactic, if not semantic, information is embedded with the modeled entity. Core CPD operations, such as requirement specification, design, costing, scheduling, construction, and operations, benefit from complete and robust modeling environments like BIM. Just-in-time access to information can cut down errors, cost, and time of completion dramatically (Akın 2008).

Finally, BIM allows the inclusion of "intelligence" in data by taking advantage of dependencies among objects as well as ability to specify different attributes of the objects modeled. This provides opportunities for providing persistent, ontological, and data-dependent structures in the building information realm. Standards for interoperability that come in many shades but with one dominant color: *Industry Foundation Classes* (IFC) provide a platform for data exchange so essential for BIM-intelligence (Akın 2008).

[10] Akın. (2008) Chapter 1: Current trends and future direction in CAD. In: Karimi H, Akıncı B (eds) CAD/GIS integration: existing and emerging solutions. Taylor and Francis, New York, NY

Digital Information Advances

Stakeholders and CPD professionals primarily, if not exclusively, rely on computer-based representations in order to design, construct, manage, and communicate. Specialized software and the development of BIM technology make the use of digital design ubiquitous. Sensor and laser technologies help gather remote and just-in-time information. Protocols of data exchange enable productive communication. Finally, building code checking applications offer more reliable and faster verification of design requirements.

The benefits expected from BIM, and in some areas substantially realized, include increased value, reduced cost, reliable delivery schedules, and improved operations and maintenance performance. The vision of such innovations to propel a new age in the AEC industry must include not just new digital tools and techniques but also advances in the practices that utilize these innovations.

We envision design and engineering practices that are in the service of sound economy, environmental conservation, improved value, greater efficiency, and satisfied occupants. CPD is no longer just about competence in the traditional sense of performing standard engineering tasks for design, construction, and deployment. Engineers and architects have to rise to the challenges of our time and ensure the best economy in providing the greatest value to stakeholders.

Performance needs to be optimized against life cycle cost of design. Value must be assessed not just for the bottom line but for other tangible (health, productivity) and intangible (comfort, happiness, recognition) benefits to the stakeholders. Last but not least, design products must meet the three pillars of market place delivery: quality, cost, and schedule.

In terms of professional and business applications, computers have come of age. Since its early beginnings in the 60s, CAD (computer-aided design) became the industry standard (Akın 2008). With the advent of BIM in the 90s, today, we enjoy almost universal acceptance of object-based representations of AEC products and processes.

IFC, an outgrowth of the *International Organization for Standardization* (ISO), has established a sound basis for universal exchange of data with reliable results. Thanks to ubiquitous internet communication, data mining and shared repositories of information are now commonplace. Special applications that can translate between different data formats, compare specifications against designs, and representations of processes and products are available for the asking.

The remaining problems of standardization, fidelity, and reliability are part of the compendium of solutions needed. These developments in the digital realm provide a rich and powerful set of opportunities for engineers and architects to address the challenges of a new and emerging world for the AEC industry and accelerating its ability to exercise increased design added value.

Exercise 1

At the end of each chapter of this text there is an exercise or two provided for the student of VBD. To start it all out, we provide a Table, below, that summarizes the basics of value assessment for buildings, which will be useful in completing these exercises.

Summary of value assessment in buildings
Key variables of value assessment and formulae for determining them

$V = p * U$; where V = value; p = probability;
U = utility; $U = B - C$; B = benefit; C = cost
UV = *Use Value* is the value of real property to occupants who use/benefit from a facility
EV = *Exchange Value* is the sale value that would be obtained by "liquidating" the property.
NPV = *Net Present Value* = Sum of all of the estimated expenditures and gains

NPV = NPV0 – Σ_[_1_–n] NPVt,
NPV0 = r0 * (*initial investment*)/(1 + (rd * *discount rate*))0
NPVt = ((rb * *benefit*) – (rc * *cost*))/(1 + (rd * *discount rate*))t

where

n = number of total years, t a given year, $1 < t < n$;
r0 = *risk of initial investment*, 0 < r0 < x;
rd = *risk of discount rate*, 0 < rd < x; rc = *risk of cost*, 0 < rc < x; rb = *risk of benefit*, 0 < rb < x;
GI = *Gain on Investment* of a built asset increases over time,
DV = *Discount Value* of features decline over time due to = *Opportunity Loss* + *Depreciation of Product* + *Operation and Upkeep* costs,
ROI = *Return on Investment;* the required rate of return + the weighted average cost of capital, after tax + the reinvestment rate.

Feel free to use this Table as a guide for developing your interpretation of value assessment in buildings. First, select a building you are familiar with and, more importantly, have access to data about its economics. You can also decide to use the case studies included in Chaps. 12 through 17. Contingent on the extent of data to which you have access, make sure you structure your calculations according to the table above.

Chapter 3
Fallingwater: A Celebrated Case of DAV Analysis

> *"My money has bought me great many fine things in life," Edgar J. Kaufmann told Wright, "but none of them have brought me greater joy than the house you built for me on Bear Run." He was speaking of course, of Fallingwater, a structure so remarkable, so romantic, so perfectly in tune and in harmony with its setting that it is understandably the most famous private residence in the world.*[1]

Edgar Kaufmann, Jr. himself was an admirer of design, and studied painting, typography, art, and architecture. Edgar Kaufmann, Jr. joined the Taliesin Fellowship in 1934, and shortly after urged his father to support and fund Wright's *Broadacre City* project, an alternative to urban centralization. To this cause, Mr. Kaufmann, his father, gave his full endorsement. During one of his trips to Pittsburgh, the Kaufmanns took Wright out to the site at Bear Run, where they owned a weekend cottage situated in a forest glen abundant with rhododendron bushes and waterfalls. At the very outset, that site so appealed to Wright's creative genius that he wrote of his first glimpse of it in December 1934: "*the visit to the waterfall in the woods stays with me and a domicile has taken vague shape in my mind to the music of the stream. When contours come you will see it.*"[2]

In fact, this shape was to be one of the most innovative conceptions of architecture where the building's natural context became a remarkable form giver and the basis of Wright's expression of a completely new set of design principles (Fig. 3.1). In a TV film on Fallingwater produced by Kent State University, Edgar Kaufmann, Jr. recollects that first trip of Wright's to Bear Run.

> *When my parents invited Wright to visit the property with the idea that he might design a house for them, they (explained that) the falls, including the drama of the water's movement and the charm of the noise that it made, were something that everybody appreciated a great deal. This was a focus for us. Eventually when Wright designed the first scheme for the house, which was also the last scheme for the house, he placed it as we all know directly*

[1] Pfeiffer in Millar CJ (1986) Frank Lloyd Wright letters to clients the press. California State University, Fresno, CA, p 82.

[2] ibid.

Ö. Akın, *Design Added Value*, https://doi.org/10.1007/978-3-030-28860-0_3

Fig. 3.1 Fallingwater's celebrated views from inside and outside. (Image credit right: "Falling Water" by Max Z is licensed with CC BY-ND 2.0)

> *above the waterfall, which I do not think had ever crossed my parent's minds* (Fig. 3.1*). But once having been presented with the notion they had no resistance to it at all, it seemed perfectly good and proper.*[3]

"Proper" may not be exactly the correct description, at least in hindsight as this placement has been the catalyst for Wright's extraordinary creation more than any other factor, as well as leading to the complications and difficulties the building went through during its stages of design, construction, and commissioning.

The contractor of the house was a local builder who, while competent in the conventional methods of the day, clearly had not done anything like this before. Some of the early stonework had to be redone, but more seriously the reinforcement work of the innovative cantilevered floor slabs were questioned time and time again and became a source of conflict for the project. Wright was confident in his design and drew from earlier experiences with the Imperial Hotel in Tokyo, which used similar technology (Akin 1993).

In the Early Years

The Kaufmanns spent their weekends in the house from 1939 on to the end of their lives. Since then Edgar, Jr. has given the house to the Western Pennsylvania Conservancy so that it might be cared for and made open to the public. Thousands visit Fallingwater each year. In fact, the Conservancy was directly involved in the repair and restoration of Fallingwater regarding two issues: moisture penetration and structural integrity.

[3] ibid.

Moisture Penetration

In the early 1980s, during one particular snowy winter of Western Pennsylvania, which is a more common occurrence that not, there were 28 different leaks in Fallingwater. The only mitigating measure taken at the time was to equip the building with equal number of buckets.[4] The architectural office of Louis Astorino and Associates was commissioned by Edgar Kaufmann to look into and resolve this issue.[5]

Most of the stone-bearing walls of the structure were designed to rise above the roofline. This was a decision dictated by the composition of the building's elevations. The horizontal ledges created by these walls and chimneys were unprotected from moisture penetration. There were no copings designed or built for these surfaces. As a result, water traveled inside the cavity of these vertical walls, eventually finding its way through skylights and wall cavities into the interior and onto wall surfaces.

The grouting seal in the spaces between stone pavers used to finish the terraces were worn out and were admitting moisture at greater rates than when first installed. The rolled finish of the bottoms of cantilevers, ends of ledges, and overhangs provided free surfaces on which precipitation ran down and into the structure or against unprotected surfaces. The roofing above these overhangs and on roof surfaces was old, cracked, and unable to keep the moisture out. Some of the skylights and windows were not well protected from moisture or simply were beaten down and worn out by the humid climate of Bear Run, Pennsylvania.

The solutions to these problems included considerable work ranging from surface repair to moisture membrane installation into structural and nonstructural elements. More specifically:

1. The top coursing of stonework was removed from all of the protruding walls. Lead moisture barriers and flashing details for chimneys were installed. The stonework was replaced to keep with the original appearance of the building.
2. Stone pavers from all terraces were removed, rubber membranes were installed. Pavers were reinstalled.
3. Rubber roofing was installed. Broken off and cracked concrete surfaces were cleaned. Loose pieces were broken off. Special round moldings were fabricated for the pressured injection of concrete finish onto these surfaces.
4. Skylights were repaired. Windows and frames were replaced as needed. Glazing was replaced with double glazing sandwiching a transparent membrane to reduce UV ray admission into the building.

In spite of all of this work, given the heavy precipitation patterns found at this region, this repair job had to be repeated later, albeit in a more limited form. In dismissing roof leakage as a problem that is unavoidable, Wright has been blamed for

[4] Personal observation of author and his guests.

[5] The issues summarized in this section are based on testimonials and documentation provided by Mr Louis Astorino, Sr.

being insensitive to the injurious nature of these problems. It is conceivable that through such comments he was aptly expressing our inevitable helplessness against nature's way with manmade structures particularly in severe climates.

Structural Problems

The sagging and cracking of the cantilevered slabs, which started during construction, became progressively more serious as years went by. Robert Silman Associates, a structural engineering firm in New York, worked on the resolution of this problem (Matteo et al. 1996).

Even though the compressive strength of the concrete used appeared to be uneven, it was not below the design stresses at 3000 lb./si. One culprit was the omission of tie reinforcing bars that connect the flow of stresses in the first floor cantilever beams to the *piloti* that support these cantilevers. In 2001, Silman measured the deflection in the cantilever beams at 5–7 inches, which was excessive.[6] The remedy developed for this problem was post-tensioning of the cantilever beams to increase their tensile and compressive strength as well as reinforce their tie into the *piloti,* anchored into the rock-bed. This approach required a strategy of preservation, leading to the removal of the flooring to reach the structure, to avoid the more difficult task of removing and replacing the original stucco from below.

A Retrospective Analysis

Fallingwater's problems were not entirely unpredictable. First of all, it was not clear to anyone outside of a very small group of individuals—namely Wright, a handful of his associates, E. Kaufmann, Jr., and possibly E. J. Kaufmann, Sr.—what these unusual design features of Fallingwater would add to the value of the building. However, the engineers who were hired by Kaufmann to verify the structural design certainly were not happy with the design. The contractor Mr. Hall was also less than enthusiastic about the challenges of this design.

Finally, there was a great deal of rancor as evidenced by Millar's (1986) publication of the correspondence between Wright and others (Table 3.1) that illustrates both the difficulties of the design and the working style of its architect. E J Kaufmann demonstrated his readiness for the challenge when he hired external engineers to double-check Wright's work, authorize the inclusion of extra reinforcing bars in the cantilever structure, and take up Wright in contentious correspondence (Table 3.1).

From the standpoint of our objectives in this text, it is worthwhile to note that this was an epic battle for the value added by Wright's unique design; a tangible value

[6] https://www.nytimes.com/2018/08/07/obituaries/robert-silman-engineer-who-saved-fallingwater-dies-at-83.html

Table 3.1 Correspondence between Wright and E. J. Kaufmann, *August 26, 1936*

August 26, 1936

Mr. Edgar J. Kaufmann
Kaufmann Department Stores
Pittsburgh, Pennsylvania

My dear E.J.:

If you are paying to have the concrete engineering done down there, there is no use whatever in our doing it here. I am willing you should take it over but I am not willing to be insulted.

So we will send no more steel diagrams. I am unaccustomed to such treatment where I have built buildings before and do not intend to put up with it now, so I am calling Bob back until we can work out something or nothing.

Also it appears that an attitude has developed on your part – "what does the architect know about what I want – I am going to live in this house – not he." Now I have heard that provincialism from women but never before from a man. And it isn't too late yet for you to get an architect that does know what you want.

I don't know what kind of architect you are familiar with but it apparently isn't the kind I think I am. You seem to know how to treat a decent one. I have put so much more into this house than you or any other client has a right to expect that if I haven't your confidence – to hell with the whole thing.

Sincerely yours,
Frank Lloyd Wright

August 28, 1936

Mr. Frank Lloyd Wright
Architect
Taliesin West, Arizona

Dear Wright:

If you have been paid to do the concrete engineering up there, there is no use whatever of our doing it down here. I am not willing to take it over as you suggest nor am I willing to be insulted....I don't know what kind of clients you are familiar with but apparently they are not the kind I think I am. You seem not to know how to treat a decent one. I have put so much confidence and enthusiasm behind this whole project in my limited way, to help the fulfillment of your efforts that if I do not have your confidence in the matter – to hell with the whole thing.

Sincerely yours, Edgar J. Kaufmann.

P.S. Now don't you think that we should stop writing letters and that you owe it to the situation to come to Pittsburgh and clear it up by getting the facts?

Sincerely yours,
Edgar J. Kaufmann
The Kaufmann Department Stores, Inc.

added to Kaufmann's pocketbook, to the legacy of Wright, the architect, and not to mention the annals of distinguished architectural edifices of the World. Without the cantilevers, in spite of their contributions both to the moisture problems and the structural issues, the value of this house would be no different from all the other house designs of Wright. While that would not have been such an insignificant recognition, compared to the "invaluable" status that Fallingwater has today, it would have been trivial. The design value of Fallingwater surprised that of a family house and contributed to spearheading a genre of design in modern and organic architecture, influencing many architects and designers to this date.

VBD Features of Fallingwater and Their "Spin" and "Buzz"

To say that the impact of Fallingwater in the field of architecture has been significant is an understatement. Arguably, the iconic American example ushered in several significant Modern design principles including the cantilevers, which incidentally worked to explode the "box," using material characteristics as decoration, and harmonizing natural and manmade resource of the site. Each one of these features of Fallingwater can be considered a VBD feature. This means that they add value through the way they were designed: the *cantilevers*, novel arrangement of physical elements such as *window placement situating mullions at corners of rooms, vertical chimney elements counterbalancing horizontal cantilevers,* and *stone ledges forming a part of the hearth of the fireplace and cascading ledges of the terraces.*

While these features were remarkable and arguably of economic value, initially, the lack of practical methods of appropriate accounting measuring-sticks prevented their accurate assessment. Therefore, they were realized only through the momentum created by the great personal stature of its architect and the equally remarkable ambition of its owner, which appears to be a VBD dream-team, made in heaven.

It is also notable that the convergence of the variety of resources that made Fallingwater possible, which may seem like an incredible coincidence, were entirely premeditated. Edgar Kaufmann, Sr., who may not have been as erudite as his son on the fine points of architectural theory, was nevertheless erudite enough about the value of a product that could add to his reputation as well as that of Wright. After all, he was a successful retailer with a deep understanding of branding and marketing. He used these skills to craft the remarkable marketing strategy for Fallingwater. Details of the evidence that supports this can be found in Franklin Toker's remarkable work entitled *Fallingwater Rising* (Toker 2003) (Table 3.2).

As a matter of sound instinct rather than hard core theory, Mr. Kauffman, Sr. asked his acquaintances and business associates to prominently feature this special building in the most prevalent medium of the day, photo magazines and journals. A remarkable coincidence that helped Kaufmann's cause is the decision by the

Table 3.2 Media features and events contributing to the promotion of Fallingwater

Evolution of publicity media	
1900s	Movies
1920s	Radio—Auditory
1936	Photo magazines—Center spreads: Visual
Design of Fallingwater	
1934	*Broadacre* project supported by E. J. Kaufmann Sr.
1935	Commissioning of a "summer cottage" in Uniontown
1935	First rendered perspective drawing at Taliesin west
1936	Completion of the design at Taliesin west
Construction of Fallingwater	
1936	Construction of cantilevers at Uniontown
1937	Completion of the building for ~ $166,000
Today	$4 M+ value; $2,5 M structural and moisture retrofit, 2000–2005
Publicity of Fallingwater—"spin and buzz" instigated by E. J. Kaufmann	
Mar 21, 1937	*Newsman Putzel*, St Louis post-dispatch the perspective spread
April 1937	*Architecture and Modern Life*, publishes three photos
July 18,1937	*Milwaukee Sentinel,* publishes article
Jan 10,1938	*Architectural Forum*, full issue with famed Bill Hedrich photo
Jan 17, 1938	The perspective on cover of *Time*; article by H. Myers, H. Luce
Jan 17, 1938	*Life* magazine
Jan 24, 1938	MoMA show on modern architecture by H. Myers, H. Luce
Feb 12, 1938	*New Yorker* article by Lewis Mumford
Feb 1938	*Pencil Points* article by Talbot Hamlin
Feb 21, 1938	*Time*, second publication, including mention of MoMA
Feb–Mar 1939	Second show at the MoMA on Wright

Toker, Franklin, Fallingwater Rising Alfred A Knopf, New York, 2003

Museum of Modern Art (MOMA) to feature the American version of Modern Architecture exemplified by F L Wright, as opposed to Bauhaus Architecture, largely due to the controversies surrounding Germany of the pre-WWII era. While both Bauhaus and the Nazi movement flourished, in their own unrelated circles, Fallingwater and Wright were the real beneficiaries of MOMA's controversial decision.

Stakeholders and the VBD Analysis of Fallingwater

VBD analysis is conducted in a multidimensional space surrounding building design features, including perspectives of several stakeholders for each feature, addressing three levels of impact—immediate, median, and long term—as well as generating estimates of value in terms of associated risks, costs, and benefits (Fig. 3.2).

In Fallingwater, there were a number of different stakeholders who were affected. These include the architect (the designer and members of his "school of thought"), the patron (owner, financier, user), the host city or region (the regional economy of the "Bear Run" area), the cultural milieu (architectural industry), other related professions (media and commercial products), and commerce (tourism and retail). In other VBD cases, there could be additional stakeholders of interest, such as the political establishment that benefits from the public interest and associated benefits and costs.

The potential benefits and harms to which these stakeholders are subjected operate within different levels of influence. One effective way of defining these levels is the time scale. For instance, in the case of Fallingwater the 1st level of impact, we call *immediate,* covers impacts during the first year after realization of VBD features. The 2nd or *median* level of impact covers the next 2–5 years of existence or reputation of the VBD featured facility. Finally, the 3rd of *long term* level of impact covers

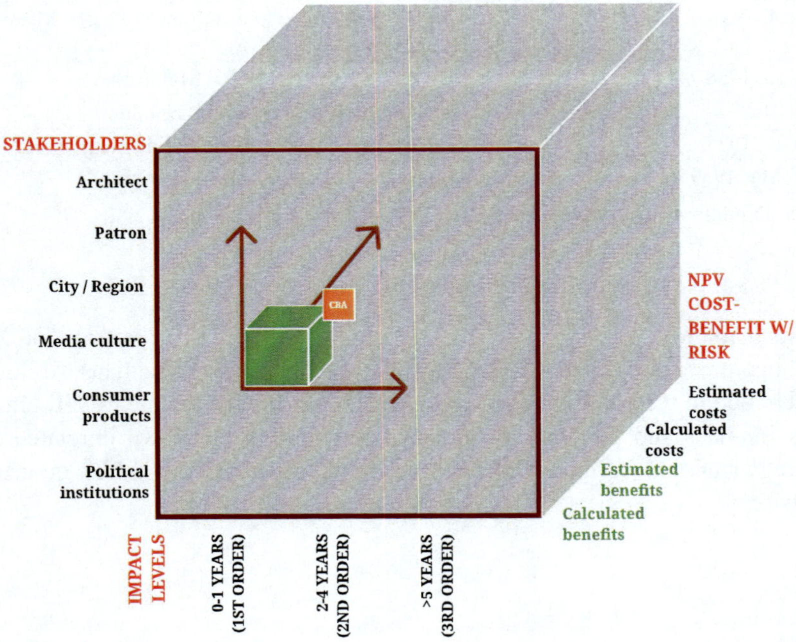

Fig. 3.2 Multidimensional Space of VBD Parameters for Each Feature, shown as the small cube

the entire lifetime of the facility. The costs, benefits, and risks in Fallingwater involve assets—personal, corporate, or environmental—that are committed; potential gains to be accrued by the stakeholders, and uncertainties that can change these expectations.

Fallingwater's Costs and Benefits with Risk

The costs, benefits, and risks in Fallingwater involve assets—personal, corporate, or environmental—that are committed; potential gains to be accrued by the stakeholders, and uncertainties that can alter these expectations. The remarkable forces that brought this building about—unusual features of its design, their immeasurable value, and the unconventional resources enabling their realization—constitute the prerequisites for and are central to VBD analysis. Below we will begin to illustrate the basic steps of such an analysis through a breakdown of the costs, benefits, risks, and mitigation measures taken in the case of Fallingwater.

The VBD features of Fallingwater represent potential benefits and costs (Chapter 4) to all of the stakeholders defined in Fig. 3.2. Benefits accrued by them may impact their finances, reputations, careers, and overall goals of professional or commercial practice. The challenge is to accurately estimate these benefits, determine if they can be attained through decisions made in the early stages of the design process, and assess whether their impact on subsequent design decisions can be measured. Similarly, we are interested in doing the same with the costs that are attributed to these VBD features.

Before developing a general-purpose method for conducting VBD analysis for all buildings—small, large, spectacular, or modest—let us begin to sketch out some preliminary answers to these questions as they apply to Fallingwater. For its design and delivery, Table 3.3 outlines some of the parameters of cost–benefit analysis, in Fallingwater, and quantitative methods of measurement associated with each parameter.

The risks that lead to the problems in Fallingwater can be traced back to the design and construction of the VBD features. Under the first order issues, since the program of the building was a modest vacation cottage and not a masterwork, Wright was fully aware that he was creating something to which Kaufmann may not be all that receptive. However, he was prepared for resistance from his patron and also to push it through, using his stature and persistence, not to mention his influence on Kaufmann's son Edgar.

Kaufmann, Sr. could not have been unaware of these risks when he decided to hire Wright. He wanted an unusual design, a signature piece to showcase his stature in business. In order to establish the proper rapport for this with his architect he went as far as subsidizing his "utopian" projects.

Wright, in his mid-60s, was by all measures a mature but aging master. His career, after having charted several glorious trajectories during the previous several decades, was in a slump. The Great Depression was in full swing and so was the depression of the building sector. Wright knew that it was a chance of a lifetime for him to take the opportunity presented by Kaufmann's patronage.

It is also not idle speculation that the design of Fallingwater was conceived as the fruit of a rich and long career by a "genius" architect, and not just a simple response to a routine commission. Consequently, Wright was looking to realize, as the 1st level of impact, his "dream project." As he knew this would reignite his career: the 2nd level of impact. Finally, the 3rd level of impact was his expectation that this commission would carry his reputation to greater heights into the foreseeable future. All of these benefits were realized (Table 3.3).

As the 1st level of impact, as a reciprocation to Wright's intentions, Kaufmann aspired to build his "dream house;" as the 2nd level of impact, he was aiming to promote his name in the media; and as the 3rd level of impact, he wanted to brand the Kauffman name, thereby promoting his family business (Table 3.3). In spite of the very high initial likelihood of these outcomes, none of them were quite realized during the time frames that are set by the VBD analysis (recall the parameters and levels presented in Table 1.3).

The environment, culture, media, and the commercial products linked to Fallingwater were factors that came about after the fact. Therefore, many of the costs such as the impact on the environment, commercialization of the area, and

Table 3.3 Some Parameters of Cost–Benefit Analysis in Fallingwater (Crouch and Wilson 1982)

	Parameters of C-B Analysis (Q.M. = quantitative measures)			
Stakeholders	Costs (three orders)	Q.M.	Benefits (three orders)	Q.M.
Architect	1st: Poor client relations 2nd: Poor reputation 3rd: Poor reputation	2* Negatives	1st: Commissions 2nd: MOMA exhibit; 3rd: Notoriety / acclaim	3* Positives
Patron	1st: Cost escalation 2nd: Usability problems, no name recognition 3rd: Maintenance costs	3* Negatives	1st: Increased property value 2nd: Select notoriety; 3rd: Tax write-off	3* Positives
Environment	1st: Roadwork 2nd: Excessive traffic; pollution 3rd: Overgrowth	3* Negatives	1st: Neutral 2nd: Tourism income 3rd: Regional center	2* Positives
Culture–Media	1st: Neutral 2nd: Architectural chauvinism 3rd: Emergence of Starchitecture	2* Negatives	1st: MOMA exhibition 2nd: Architecture publications 3rd: Global acclaim; west-Penn Conservancy	3* Positives
Commercial products	1st: Neutral 2nd: Commerciali-zation of Fallingwater. 3rd: Commerciali-zation of architecture.	2* Negatives	1st: Neutral 2nd: Income; copycat commercialization 3rd: International awareness	2* Positives

Crouch, E. A. C. and Wilson R. Risk/Benefit Analysis Ballinger Publishing Co., Cambridge MA 1982 potential architectural advances were nonfactors at the time of the initial stages of design and delivery of the building.

Finally, the myriad of the stakeholders, whether they were in the publishing business, or the museum sector, grew their patronage through the exposition of Fallingwater's unusual features and added to the snowballing of the building's reputation. An irony, which is aptly noted by Toker (2003), is the name of the building, "Fallingwater," coined by one of the reporters during the early days of its construction. Much to E.J. Kaufmann's chagrin, by making no reference to name, unwittingly, the reporter denied him publicity benefits he expected to have.

It is also worth noting that these assessments are *post facto* and do not help the design decision-making process at the time these VBD features were conceived and covered by the delivery schedule and budget. To make a real impact on the potential VBD success of a building, the costs and benefits need to be assessed *pre facto* at the time of requirement specification and architectural programming phases (Table 3.4).

Benchmarking the Costs and Benefits of Fallingwater's Design Features with Risk

The remarkable forces that brought this building about—namely, its unusual design *features*, immeasurable *values* associated with these features, and the unconventional *resources* used to enable their realization—constitute the prerequisites for the VBD in this project. Let us illustrate the pre facto case through a breakdown of the costs, benefits, and risks, at the time of initial design, without the benefit of 20/20 hindsight.

During the design of Fallingwater the inherent value of cantilevers was a point of contention. In the annals of Modern Architecture, this was such a novel feature at the time that even its costs were difficult to estimate, let alone its benefits. In such cases benchmarking is absolutely invaluable.

Benchmarking can create reasonable estimates for cost and benefit based on previous and similar cases. Then it becomes possible to make these estimates more accurate through a risk-coefficient that expresses the level of confidence we are willing to attach to them. In Fallingwater, we see a form of this in Wright's correspondence with his client E. J. Kauffman (Table 3.4). Wright is benchmarking based on his previous experience of handling challenging design features in his other buildings, not necessarily based on any comparable design features as to the cantilevers in question. Proxy experiences in the absence of close comparable examples of benchmarking are still better than none.

The VBD features of Fallingwater during the design phase represented potential benefits and costs to all of the stakeholders defined in Table 3.3. Benefits accrued by these stakeholders could have impacted their finances, reputation, careers, and overall goals of practice. The challenge was to be able to accurately estimate the risk

Table 3.4 F L Wright correspondence referring to earlier works as evidence of pre facto benchmarking for Fallingwater (Millar 1986, pp. 97–98)

August 29, 1936
Mr. Walter J. Hall
Kaufmann's Bear Run Camp
Pittsburgh, Pennsylvania

My dear Hall:

I guess I took too much for granted when I called you on the Kaufmann house. Probably you have always been your own boss, never worked for an architect and never had ethics.

At any rate it appears from Bob's letters that you are undertaking to advise Kaufmann concerning various matters that are certainly my business and none of yours, and this tardy letter from you shows that it is by no accident on your part but is your habit of mind.

What proves this to me is your reference to Bob "not getting any money out of the job," while you Kaufmann and I are – say you.

Well if Bob is any good he is getting far more money out of it. In any case I would save you further anxiety on his behalf by reminding you that it is exactly none of your damned business no matter which way you may look at it. As for myself, I shall get so much less than money out of this house that I prefer not to be mentioned in that connection with Kaufmann and yourself.

When I sent for you I hoped to find real help in getting a difficult building built for a difficult client. But it seems to me now I sent for you only to find I have greater difficulty than either client or building in the man picked to be my right-banner. I saw "back down the line" as I read your brief and tardy and impertinent letter. As I learn what has happened to our plans and details as have been sent to Bob I see that I am in trouble.

I have therefore called Bob back here until some basis for mutual respect between myself and Mr. Kaufmann is reestablished.

You are now included in this armistice if such it proves to be.

If you imagine your meddlesome attitude to be either sensible or honest (we will not say ethical) something was left out of either your character or your education.

I have put too much into this house (even *money,* which item you will understand) to have it miscarry by mischievous interferences of any sort. The kind of buildings I build don't happen that way. Several have been ruined that way. And this one may be one of them.

It is only fair to say to you directly that you will either fish or cut bait or I will. I am willing to quit, if I must, but unwilling to go with my eyes open into the failure of my work.

With a setup such as the present one turns out to be, there can be only failure. I have built one hundred and ninety of the world's important buildings without knowing the look of the thing when it turns up on the job.

Failure, I mean, by way of treacherous interference,

Sincerely yours,
Frank Lloyd Wright, Architect

these expected benefits and costs were subject to determination of whether they were worth the investment, based on associated risks, and whether their impact on subsequent design decisions would be measurable.[7]

Similarly, we would be interested in doing the same analysis with the costs of these design features. The values suggested for each of the cost–benefit estimates (Table 3.5) are at best ordinal predictions based on experience. Regardless of how good our intuition for predicting these values may be, we need to measure them quantitatively. This is at least possible in the post facto approach. Table 3.5 shows values that are obtained from appropriate benchmarks, actual statistical values as well as methods of measuring values for all five stakeholders for Fallingwater, in three levels of impact.

We will devote greater attention to the subject of benchmarking in the following chapters.

Exercise 2

A unique feature is what adds value to the design of a building. A *benchmark* is a comparative baseline for assessing differential value assessments of that design feature. Develop feature and benchmarking descriptions for a building of your choice where you have sufficient information to draft a table like the one provided below.

Example: Feature and benchmarking descriptions in fallingwater				
#	Feature	Description	Benchmark	Description
1.	*Cantilever structure*	Three floors of 10-17 feet unsupported extension of floor-slab over the waterfall	*Load bearing structure*	Perimeter walls and interior load bearing elements are carried straight to foundation
2.	*Corner mullions*	Novel elements: place window mullions at corners of rooms	*Regular mullions*	Windows are placed in the body of walls not corners
3.	*Vertical elements*	Chimney and walls counterbalancing hoizontal cantilevers	*Walls and chimneys*	Walls and chimneys are placed in relation to the internal functions
4.	*Natural elements*	Stone ledges forming part of the fireplace hearth and the terraces	*Placement on site*	The building is placed across from the falls on bear run creek

[7] Schoemaker, P J H (1980) Experiments on Decisions Under Risk: The Expected Utility Hypothesis, Kluwer-Nijhoff Publishing, Boston.

Table 3.5 Cost–benefit metrics for fallingwater

Stakeholders	Parameters of pre facto C-BR analysis			
	Costs	Some metrics	Benefits	Some metrics
Architect	• 1st: Poor client relations • 2nd: Poor reputation • 3rd: Poor reputation	• 1/16 bad to good ratio • Compromise 16 times • Lose 16 clients for 1 compromise	• 1st: Commissions • 2nd: MOMA exhibit • 3rd: Notoriety/acclaim	• $2100 in 1937 • ½ * $40,000 first impact • ½ * 340 (532–192) client
Patron	• 1st: Cost escalation • 2nd: Usability problems, no name recognition • 3rd: Maintenance costs	• 30 K: 250 K (1:8.33) • Loss of commercial impact ~0 • $5,750,000/75 years +/~ $100,000/year	• 1st: Increased property value • 2nd: Select notoriety • 3rd: Tax write-off	• Inflation, • Gift to Western Pennsylvania Conservancy (tax write-off) ~ %20 of $800,000 • 2003 renovation $11.5 million ~ $20 M value
Environment	• 1st: Roadwork • 2nd: Excessive traffic; pollution • 3rd: Overgrowth	• 7700 kg of Co2 per year per car • $60,000 per lane per mile • Tourism ~2KW/$	• 1st: Neutral • 2nd: Tourism income • 3rd: Regional center	• Heritage tourism in PA: 32million visit = $1billion • 165 sites 67 in Philly • $6 million per site
Culture–Media	• 1st: Neutral • 2nd: Architectural chauvinism • 3rd: Emergence of Starchitecture	• Burden of master architect ~1:2 cost overruns and delays, • Reputation of all architects	• 1st: MOMA exhibition • 2nd: Architecture publications • 3rd: Global acclaim; west-Penn conservancy	• What would you do to estimate this? [*this exercise is left up to the reader to figure out*]
Commercial products	• 1st: Neutral • 2nd: Commercialization of Fallingwater • 3rd: Commercialization of architectural	• What would you do to estimate this? (*this exercise is left up to the reader to figure out*)	• 1st: Neutral • 2nd: Income; copycat commercialization • 3rd: International awareness	• Fee: $6 ~ $60 • Average value of museum items $34.86 ($14.95–$234.00)—*eMuseumStore.com*

Chapter 4
Value-Based Design

Real-Estate Economics

Economics is one of the central disciplines that have paved the way to decision-making in general and *rational decision-making* in specific. In turn, decision-making is the skill of choosing between alternatives. The three basic ingredients of decision-making are: (1) the *decision maker*, (2) available *alternatives*, and (3) the resulting *choice*. The decision maker is the agent that ultimately makes the choice. Alternatives are the options that are available to the decision maker. There are always alternatives in any decision-making case, including the choice to not choose. The ultimate result of a decision-making case is the choice made using these specific methods.

Consider the well-known economic choice problem: "would you prefer a raffle ticket that (A) guarantees a $50 payoff, or (B) one that yields $495, with 1 in 10 chance of happening?" If you are a follower of the *Rational Choice Theory* (RCT), you would pick alternative A. According to this theory the net value of a choice is equal to the product of the probability and value of the available option. Choice *A* yields a payoff of $50 while choice *B* on the average yields only $49.50.

The trouble is RCT is not infallible either. Its weakness is demonstrated when we take the choices that are clearly distinct, like the one above, and convert them into *equivalent* ones: "would you prefer a raffle ticket that (A) guarantees a $50 payoff, or (B) one that yields $500, with 1 in 10 chance of happening?" RCT states that alternatives *A* and *B* are equivalent and decision makers should be indifferent in this choice.

However, in reality everyday decision makers have no difficulty choosing. Some prefer *A*, because it is a sure bet and the payoff is just right for a useful end. They think just winning something is so novel that they cannot reject a sure bet. Others are taken by the thrill of trading in a sure bet for a chance at a value ten times the sure bet.

Ö. Akın, *Design Added Value*, https://doi.org/10.1007/978-3-030-28860-0_4

Qualitative Estimates of Investment Value

When an entity, an individual or institution, decides to invest in a product, there is an expectation that there will be economic gain, or *Return on Investment (ROI)*. This value is the difference between the investment value and appreciated value and is called the *Net Present Value* (NPV). Figure 4.1 shows NPV for a given commodity over time. The intercept of the linear function is the value of initial investment (I) and its slope is the rate of return.

With minor modifications, the economic model shown in Fig. 4.1 can be applied to investments in buildings. The ROI can be estimated in alternative ways. It can be based on:

1. The required rate of return.
2. The weighted average cost of capital (after tax), to indicate the rate of return against the capital assets held to satisfy outstanding loans, or.
3. The reinvestment rate.

Since buildings are projects that can take years, sometime decades, to complete or to update to comply with changing conditions, the model in Fig. 4.1 must be modified to accommodate financial factors that evolve over time, such as reinvestment, reuse, or rehabilitation. NPV calculation using variable discount rates, if they are known for the duration of the investment, better reflects the real situation than

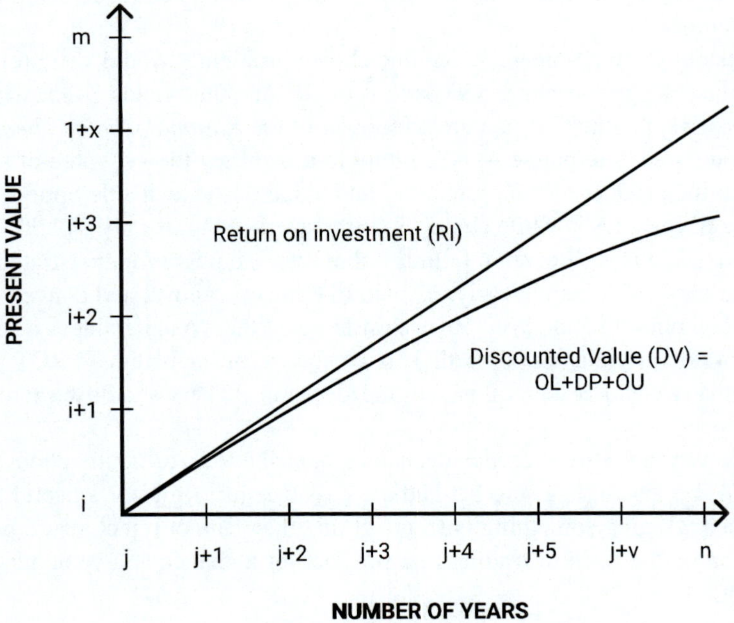

Fig. 4.1 Net present value of investment (Bon 1989) (where *DV* = Discounted Value; *OL* = Opportunity Loss; *DP* = Depreciation of Product; *OU* = Operation and Upkeep)

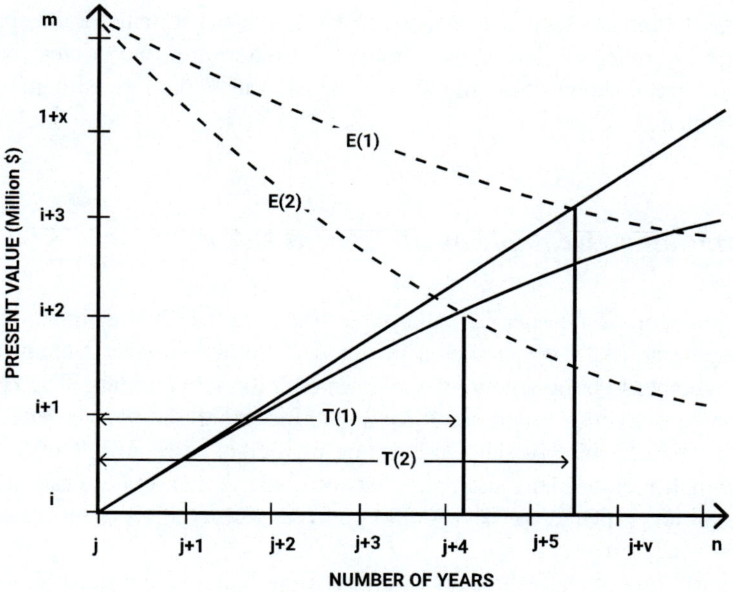

Fig. 4.2 Present value of building investment (Bon 1989)

one calculated from a constant discount rate for the entire investment duration. Thus, the economic value of buildings can be computed on the basis of the two quantifiable factors that are also used for other products: the ROI and DV (Fig. 4.2).

Over time, an investment loses value due to a number of factors. When this happens the value of the product must be discounted. There are multiple factors that impact the true value of a product through the DV. One is *Opportunity Loss*, which is the rate at which the capital needed for the project could return, if invested in an alternative venture. A second one is the *Depreciation of Product* due to aging. A third one is the *Operation and Upkeep* costs of a product (Fig. 4.2).

Qualitative Estimates of Building Investment Value

The value of real estate is assessed as the value of alternative choices through the interaction of the NPV, taking into account the ROI and DV (Bon 1989). As the Gain of Investment (GI) of a built asset increases over time, the DV of alternative design features would decline over the same period of time. Where these two curves intersect marks the point of indifference (Fig. 4.2).

Bon (1989) shows how this can be done to assess the decision to implement phased investments. The constructed portion of a capital project is subject to DV (curved, solid line, in Fig. 4.2), while the addition or improvement being considered for future implementation shows how it is discounted over time (for T1 or T2), starting with the current time. Allowance is made for the possibility of changes in the economic factors and two different rates for DV, one an estimate for the upper bound (E1) and the other for the lower bound (E2). Where the two curves (broken and solid) intersect shows the time before which the added investment should be implemented.

A Quantitative Example of Investment Value

Let us now consider a specific example to illustrate the NPV estimate in a phased capital project, like the one shown in Fig. 4.2. Let us consider a client who must decide whether to build a new feature as an addition to a building. The new feature will have startup *costs*, operational *costs*, and incoming *cash flows* (benefits) over 6 years (t = time). This project will have an immediate cash outflow of \$700,000, at $t = 0$, which includes soft and hard costs of construction. Other cash outflows for years 1–6 are expected to be \$50,000 per year. Cash inflows are expected to be \$100,000 each for years 1–6.[1]

All cash flows are after-tax, and, for after year 6, there are no cash flow estimates available. Let us also assume that the required rate of return is 10% and a risk factor for exposure over the initial investment, added cost, added benefit, and discount rate estimates to be 1.0, while in reality each component is likely to have different risk rate. Given our assumptions, the NPV can be calculated using the following formulation:

$NPV_0 = NPV_0 - \sum [1-n]\ NPV_t,$
$NPV_0 = r_0 * (\text{initial investment})/(1+ (r_d * \text{discount rate}))^0$
$NPV_k = ((r_b * \text{benefit}) - (r_c * \text{cost}))/(1 + (r_d * \text{discount rate}))^t$where
n = the number of total years, t is a given year, and $1 < t < n$;
r_0 = risk of initial investment, $0 < r_0 < x$;
r_d = risk of discount rate, $0 < r_d < x$;
r_c = risk of cost, $0 < r_c < x$;
r_b = risk of benefit, $0 < r_b < x$;

In Table 4.1, the sum of all of the Present Values (PV) estimated for annual intervals constitute the NPV. In this example, the NPV equals −\$395,164. Since this is less than zero, unless there is an alternative investment project with a higher NPV, it would be a better not to invest in the project or extend the NPV period to one where there would be a predictable positive NPV. More complex problems would need to consider other factors, generally including the calculation of taxes, uneven cash flows, and salvage values (sale value after depreciation), in addition to the availability of investment opportunities.

[1] http://www.absoluteastronomy.com/topics/Economics

Table 4.1 NPV calculation example (https://www.uclan.ac.uk/staff_profiles/jacinta-c-nwachukwuphp)

Design feature	Year	PVs (expenditure + gain)	Yearly NPV
Expenditures	$t = 0$	$-700{,}000/(1 + 0.10)^0$	−$700,000
Annual gains over six years	$t = 1$	$(100{,}000 - 30{,}000)/(1 + 0.10)^1$	$63,636
	$t = 2$	$(100{,}000 - 30{,}000)/(1 + 0.10)^2$	$57,851
	$t = 3$	$(100{,}000 - 30{,}000)/(1 + 0.10)^3$	$52,592
	$t = 4$	$(100{,}000 - 30{,}000)/(1 + 0.10)^4$	$47,810
	$t = 5$	$(100{,}000 - 30{,}000)/(1 + 0.10)^5$	$43,462
	$T = 6$	$(100{,}000 - 30{,}000)/(1 + 0.10)^6$	$39,513
	Total NPV for 6 years		−$395,164

Qualitative Estimate of Value Added Investment

When we consider design features as value adding investments, rather than ordinary investments, in a capital project, Bon's NPV calculation can be adapted to VBD analysis with a few modifications. Both DV and GI apply to value adding design features. These features are subject to depreciation and maintenance costs as any other real commodity. However, there is one major difference; by definition the value adding design features must appreciate over time rather than depreciate. Thus, there must be allowances made for value volatility, which we call VBD *Appreciation*.

In this case, DV is the value of the deduction that needs to be made to the initial value of a feature due to aging and maintenance that depreciate its value over time. GI is the profit that is expected in the market place, if the funds invested to realize the value-adding design feature was made elsewhere. Both DV and GI are quadratic functions decreasing in relation to the investment made towards the design feature (Fig. 4.3).

For instance, deciding to make an investment in a VBD feature can present the following decision alternatives:

1. Spending $X to build a facility without the features where the discount value is −$x and the gain on investment +$x, versus.
2. Spending $Y to build a facility with the features where the discount value is -$y and the gain on investment +$y. This requires a "qualitative economics" model to manage the differential $Y-$X. This model is be based on the term Σ (−$x, −$y, +$x, +$y).

This application to VBD analysis brings up the following important considerations:

- Volatility of economic utility estimates of buildings.
- The rationale of design value.
- Research of estimates necessary for applying economic analysis to VBD.

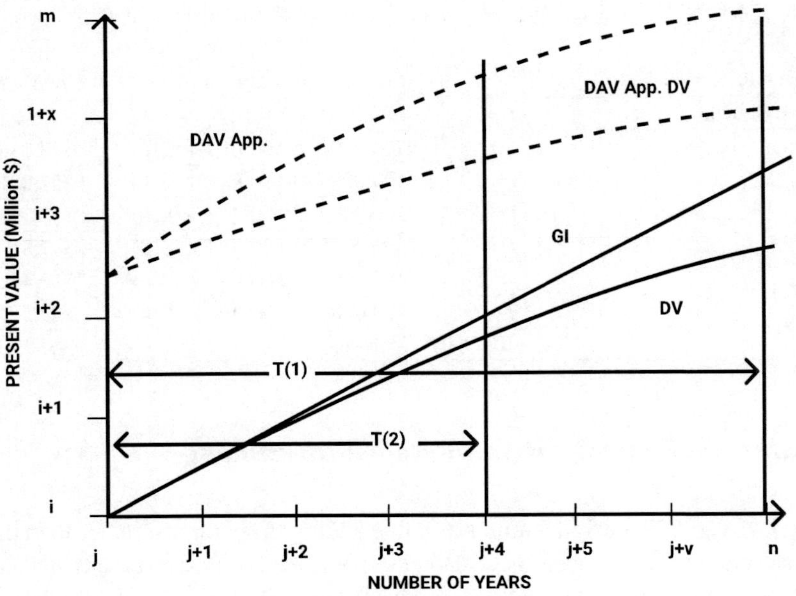

Fig. 4.3 Net present value of VBD investments (Bon 1989)

The estimates of ROI, DV, and VBD Appreciation can be treated as decisions of the *Rational Choice Theory* type. The likelihood of creating a VBD feature (p) times its economic utility to the project (U), which has two major components, cost (C) and benefit (B). This yields the total *Value Added* (VA) by a design feature:

VA = $p * U$, where $U = B - C$.

Use Value Versus Exchange Value

Use Value in VBD is the value of real property to occupants who benefit from a facility. This is ordinarily measured as the rental cost of equivalent facilities in the market place. However, this is not always the most accurate way of estimating *Use Value*, especially for value-adding design features that have branding and *prestige value*, as in the case of headquarter buildings and novel structures.

A building like Fallingwater's Use Value is difficult to equate even to a rental property of equal size and quality. Its real use is the "museum" function it serves today for the connoisseur of architecture, which was not its envisioned value at its time of design and construction. Therefore, the revenues it brings as a site of visitation and other benefits both tangible and intangible surrounding its museum function must be taken into account. Furthermore, this Use Value does not necessarily match its *Exchange Value* which is the real sale value that would be obtained upon the "liquidation" of the property.

In the previous chapter, where we introduced Fallingwater, we reported on estimates of the costs and benefits of the property. If the owners were hard pressed to sell and the buyers were value-minded entrepreneurs, it probably would go for over 20 million dollars. A connoisseur looking to benefit in ways outside of normal economic terms may well spend even more than that.

Therefore, the estimation of these values for VBD cases is a complicated exercise, to say the least. It must take into account the variety of preferences assigned by its users to the quantity, the anticipated and unanticipated uses of the building, as well as quality of the features present in buildings. Ultimately, in the hierarchy of human needs and the goods that satisfy them, buildings are one of the most complex and high-order objects that they possess.

Pre Facto Versus Post Facto Analysis

A central distinction in VBD analysis is *pre facto* versus *post facto* assessment of value. The former refers to estimates of value before a project is realized. This is invaluable for VBD since it helps predict the value of a potential design feature before it is constructed. Pre facto analysis is a form of risk assessment to see if the investment is warranted given a certain value prediction for that feature.

Post facto analysis, on the other hand, is the measurement of the costs and benefits accrued after the fact, after a project is realized. This is critical in verifying the success of the predicted VBD; but more importantly it provides benchmarking, without which pre facto predictions would not be possible.

As we go through specific methods of VBD in the following chapters, when applicable, we will point out distinctions that apply to these methods from the differing vantage points of pre and post facto analyses.

Exercise 3

Value Engineering is a time tested and frequently used method to evaluate the efficacy of cost of buildings in the budgeting phase. The purpose of value engineering is to save money and avoid features that only amount to gold plating.

Comparison of value engineering and value-based design	
Value engineering	Value-based design
During pre-construction conception phase	
Value resides in the flexible functions	*Value* resides in the flexible features
Value resides in the selling functions	*Value* resides in the *Selling Features*[xxxvii]
Isolate and minimize costs of flexible and selling items	Include and maximize flexible and Selling Features

Comparison of value engineering and value-based design	
Value engineering	Value-based design
Look for Selling Features that can be eliminated	Look for the *value* of selling designs
Look for flexible functions that can be eliminated	Look for the solution *values* of flexible functions
During post-construction evaluation phase	
Select complete functional sets	Select corresponding design sets
Add costs for each set (RCB)	Add costs for each set (RCB)
Disadvantages (breakdown by fixed/variablc cost)	Advantages (benefits, lifecycle tradeoffs)
Associate costs and benefits	Consider value based design

Use the table of comparison between value engineering and value-based design, above, to validate the principles indicated in it, based on information you have about a specific building that can confirm or refute its assertions.

Part II
DAV Analysis Methods

The method *par excellence* for Value-Based Design (VBD) is Design Added Value (DAV) analysis, which we will cover in detail, in the next chapters. First, we will review the fundamentals of cost–benefit analysis with risk methods. This will be followed by methods that constitute the basics of the DAV approach, such as data elicitation, pre facto and post facto analysis, scaling ordinal values, risk analysis, Bayesian probability, sensitivity analysis, three optimization techniques, and tenets of expertise-innovation-and-creativity in design.

Chapter 5
Cost–Benefit with Risk Analysis

Depending on the specifics of the Capital Project Delivery (CPD) process followed as well as the stakeholders and sources of information available to design professionals, there are a myriad of methods that are available in calculating the DAV of given design features. The central methodology we recommend for this purpose is cost–benefit analysis with risk. In this chapter, our mission is to describe specific techniques to implement classic cost–benefit analysis methods that can be effective in analyzing DAV features.

DAV analysis begins with identifying the potentially value adding *features* of a capital project, whether a building or an infrastructure facility. Next, the *stakeholders* who would potentially benefit from the realization of these features would be enumerated as we exemplified in Chap. 2. Stakeholder preferences drive design added value; therefore the costs and benefits of realizing any design features would be estimated around stakeholder preferences. These costs and benefits are subject to risk that needs to be assessed particularly if this is a pre facto analysis (*during the design stage*) rather than a post facto analysis (*after construction*). And finally, these benefits and costs would be split between three time periods: *immediate, median-term*, and *long-term*.

Cost–Benefit Analysis in the DAV

All design features entail some cost, either in their inception, construction, or use. We will break these down into *Soft Costs, Construction and Operating Costs,* and *Other Costs*.

Ö. Akın, *Design Added Value*, https://doi.org/10.1007/978-3-030-28860-0_5

Soft Costs

Soft costs include the cost of hiring architects and engineers to complete the delivery of a building or infrastructure project. This can be on a standard fee basis (in the USA, 5–10% of total construction cost) with considerable variance depending on the building type, client, site difficulties, and stature of architect.

Stone (1983) distinguishes between buildings of *production*—such as hospital, plant, office, and commercial—versus those of *consumption*—such as church, house, and monument—with some building types remaining *in between*—such as research facility, government facility, and school. Stone also defines *custom designed* buildings like headquarter buildings, monuments, and villas, as opposed to *standard buildings* like post office, fast food, gas station, and condominium. The DAV analysis dimensions applicable to these building types (Table 5.1) are those seeking environmental benefits which include *stature, productivity*, and *technological benefits*.

Clients are motivated by different goals for different *building types*. In the case of E. J. Kauffman (Chap. 3), for example, we see commercial and personal goals of stature overriding other goals. In comparison to this, for example university buildings combine stature with environmental benefit, when the president of the university declares that all buildings on the campus will be LEED certified. For example, Carnegie Mellon University located in Pittsburgh, PA, took this posture, accepting the required additional work and attention to green building design issues.[1] This can increase soft costs as would be the case in custom and production type buildings (Stone 1983).

Some building *sites* are more demanding than others. Gates-Hillman complex on the Carnegie Mellon University campus comes to mind. In this case, not only did the designs turned out to be more complex than originally intended, but also there were significant schedule and cost escalations related to moisture features and steep contours found on its site. Consequently, design work and associated soft costs turned out to be higher than expected. Finally, the *statue of the architect* is a significant determining factor for soft costs. Ordinarily, name architects require substantially more fee than others.

Construction and Operating Costs

Construction and operation costs include a myriad of items related to materials, installation, and overhead (Table 5.2). First costs are the cumulative of all of the items to the right of that box in Table 5.2. It includes all that is expended prior to occupancy. Operating costs, or life cycle costs, are usually related to HVAC, lighting, indoor air quality, and productivity factors. Since operating costs is a significant

[1] Based on an interview conducted with Gerald Cohen, President of CMU, in 2006, the guesstimated value of this is around 1 million dollars.

Table 5.1 DAV dimensions: stature, productivity, and environment

Building Type	Production	Consumption	Both
Custom	Stature productivity environment	Stature environment	Stature (environment)
Standard	Stature productivity environment	Productivity environment	Productivity (environment)

Table 5.2 Cost categories and their mapping in facilities (Mudge 1989)

<table>
<tr><td rowspan="10">Total cost (life-time cost)</td><td rowspan="8">First cost (selling price)</td><td colspan="5">Profit</td></tr>
<tr><td colspan="4">Selling cost</td><td rowspan="7">Cost of goods</td></tr>
<tr><td colspan="4">General administration costs</td></tr>
<tr><td>Field service and Miscellaneous</td><td rowspan="3">Overhead costs</td><td rowspan="4">Conversion costs</td><td rowspan="5">Cost of manufacturing goods</td></tr>
<tr><td>Indirect labor</td></tr>
<tr><td>Indirect material</td></tr>
<tr><td>Direct labor</td><td rowspan="2">Prime costs</td></tr>
<tr><td>Direct material</td><td>Material costs</td></tr>
<tr><td rowspan="2">Use cost</td><td colspan="5">Maintenance costs</td></tr>
<tr><td colspan="5">Operating costs—shipping, receiving, installation, etc</td></tr>
</table>

component in the life of a building, we treat it as a separate item in cost-benefit analysis. Utility costs, maintenance costs, liabilities and insurance costs, and promotional costs are part of the operating cost category for DAV analysis.

Costs of capital projects are well documented but these are generally confidential records that holders of this information consider private or proprietary. Also, sometimes the figures published or made available can be difficult to classify into the categories enumerated in Table 5.2. Interpolation of these numbers and projections based on benchmarks obtained from other capital project cases may be necessary to obtain accurate values.

Other Costs

In addition to standard expenditure items, there are other potential costs to consider in the DAV feature analysis. A client rejecting a preliminary design is a lot less costly event than a 56-story building collapsing. A missed deadline can be corrected, just as a rejected design can be, with additional work, but the un-evaluated, faulty design or undervalued design presents a fundamental problem for the entire architectural profession. The cost of fixing each of these design issues requires very different resources and the probabilities associated with each design item must be taken into account. It is important to recognize that the cost of designing and constructing novel design features means that one has to reduce the associated cost–risk combination to acceptable levels. A 50–50 chance on gaining the client's approval may just be acceptable for most architects and engineers. A VBD feature can increase these odds considerably.

An experienced designer learns from each design experience and therefore progressively improves the chances of positive results of realizing VBD features. Years of experience represents a cost to the architect as well as the bottom line of the project's budget. The cost of a delayed project and collapsed building can be more accurately estimated, although estimating the loss of life in monetary terms can invite fierce debate and disagreement. While the former represents an insignificant fraction of the latter, there are cases in which the two factors are interconnected: time-pressure leading to fatal design errors, detection of failure delaying a project for years. While there is no obvious solution strategy for any given client, it is safe to assume that the odds improve with the architect's years of experience, self-selection of clients, and improvements in the methods of the CPD sector.

Benefits

Without an understanding of the benefits in a given case there can be no VBD decision-making. Each feature that deserves the investment of time and resources required to realize it must be assessed for the potential VBD outcomes that it will offer. These include benefits to stakeholders like professionals, patrons, local institutions, culture, media, commercial establishments, and political institutions. More tangibly, these benefits can mean getting new commissions, being invited to exhibitions, publications, improved property value, improved reputation, improved tourism, new investments, and the appreciation of nearby properties.

These are important factors for DAV analysis. However, their measurement in terms of qualitative metrics, such as reputation, user satisfaction, and publicity, is difficult. Even quantitative ones, such as appreciation of market value, sales receipts, number of patrons, number of new jobs, and NPV, are not all that easy to calculate. Often the data necessary to make these estimates are not accurately recorded. Even when they are, there are difficulties of accessing the original data sources. More often than not, creative information elicitation techniques must be used to make reliable estimates, which will be discussed in Chap. 6.

VBD projects are likely to involve one or more innovative features. When these features are conceived, they often run into the potential difficulty of being rejected by the project owners or managers. As with all innovative features, there are risks of failure. Investors and capital project assessors, who control investment decisions by financial institutions, are reluctant to approve unproven features. In cases when

VBD features are cleared for implementation, the benefits may not be immediately attainable. To get them implemented extraordinary investment of resources, often disproportionate with short-term payoffs, may be required.

Risk analysis (Chaps. 7 and 8) can provide an estimate of the likelihood of implementation and expected ROI. These expected costs and benefits and the probabilities of realizing them provide the essential parameters that must be taken into account while making VBD decisions, as in deciding to build, estimating completion time, and requesting fees for services.

For example, in the K. C. Hyatt Regency case (Chap. 13), there is a trade-off between meeting the deadline while saving from the soft cost budget against the risk of collapse, however small. In that case, there were tangible consequences that could have been, and should have been, measured and included in decision-making. Risk of collapse; consequently, risking safety of occupants, must be a de facto risk analysis item for any building project.

There are a handful of commonly known good professional practice strategies that can insure reliable deadlines, such as better communication, contractual protocols, legal accountability, proactive planning, contingency planning, consultant screening, lean construction, integrated design process, and sound scheduling practices. The costs and benefits of these measures must be commensurate so that the reduction of risk would not only be desirable but also feasible.

In classical cost–benefit analysis, the normative formulation of value is:

$$\left\{\alpha(B,t)B(t)-\beta(C,t)C(t)\right\}; \tag{5.1}$$

where B is benefit, C is cost, at time t, with α and β as coefficients of conversion into a common unit, for benefit and cost, respectively, for a given design feature and stakeholder.

Similar to the formulation in (5.1), based on the cost and benefit of a situation we can represent the value of a design decision-making strategy with risk, as:

$$\left\{\alpha(B,t)B(t)-\beta(C,t)C(t)-\gamma(R,t)R(t)\right\}; \tag{5.2}$$

where t is time; and α, β, and γ are coefficients of conversion into a common unit, for benefit, cost, and risk, respectively, for a given design feature and benefiting stakeholder.

In this case the cost-benefit analysis aims to:

$$\max\left\{\Sigma_{k,l}\left[\Sigma_{j,i}\left(B_d-C_d\right)R_d\right]\right\} \tag{5.3}$$

where B_d are the benefits subject to risk R of a decision d, C_d are the cost subject to the same risks, and R_d are the risk factor involved in a decision d, k and l define the range of costs and i and j the range of benefits involved with the decisions.

Thus, a decision package tries to improve the total gain achieved by the benefits after the costs are deducted, both subject to risks. This formulation makes an obvious assumption rarely available in complex decision making situations: the scalability of all risks, costs, and benefits along the same measurement dimension. The difficulty of this proposition is that, the risk of losing one's reputation for punctuality cannot be measured in the same terms (units) as likelihood of paying a parking ticket (an item of measurable currency). Frustration of leaving a meeting in which a client is dissatisfied is very different from the disappointment of payments withheld due to change orders. Likewise, the cost of human life for insurance companies is not the same as the loss of a loved one or even the death of total strangers. At a minimum, the formulation in (5.3) must be amended:

$$\max\left\{\Sigma_j\left[\Sigma_l\left(\alpha B_d-\beta C_d\right)\phi R_d\right]\right\} \tag{5.4}$$

where α, β, and Φ are correction coefficients to map B_d, C_d, and R_d into a common cardinal scale.

In reality, of course this remedy is purely a theoretical one that makes an allowance for the possibility of achieving such a mapping rather than insuring that such a mapping is always possible. In the end, in order to support a specific decision that takes into account potentially negative or positive consequences of an action. The general strategies in completing a cost-benefit analysis include, but are not limited to, *zero-risk*, *as-low-as-reasonably-achievable* (ALARA), *best-available-control-technology* (BACT), or *full cost-benefit analysis*.

Zero-risk analysis is suitable when there are no options that are likely to lead to a proposition K, for which $p(K) > 0$. These are events that are commonly considered to be "impossible," such as the sun not rising, gravity not attracting, or taxes not being collected.

ALARA is defined as the risks involved in all "reasonable" occurrences, such as designing a span within "normal" physical and monetary limits, arriving at your flight destination, or having lunch.

BACT applies to those cases where the risks are governed by the "best" option that is commercially available at reasonable cost. Options outside of these are subject to full cost-benefit analysis where one has to explicitly account the values for risks, costs, and benefits.

The cost of the building collapse risk is so high that regardless of the feasibility of risk reduction, i.e., *risk averse behavior*, (Fig. 5.1), it must be achieved at any cost. It also turns out that the entire discipline of engineering has been geared towards this goal. Building codes, professional licensing procedures, design safety protocols, and a myriad of checks and balances established in design delivery are all to make sure that such risks never rise above acceptable levels.

There is a simple way of measuring risk: ask a significant number of members of a relevant population. "Do people consider life in the year 2000, for example, to be more or less risky than life in 1980s or 1960s?" Marsh and McLennan Companies (1980), in fact, did this with a sample population, the exact years in our example notwithstanding. They found that corporate executives considered 1980s about as risky as 1960s. Federal regulators, members of congress, investors-lenders, and the

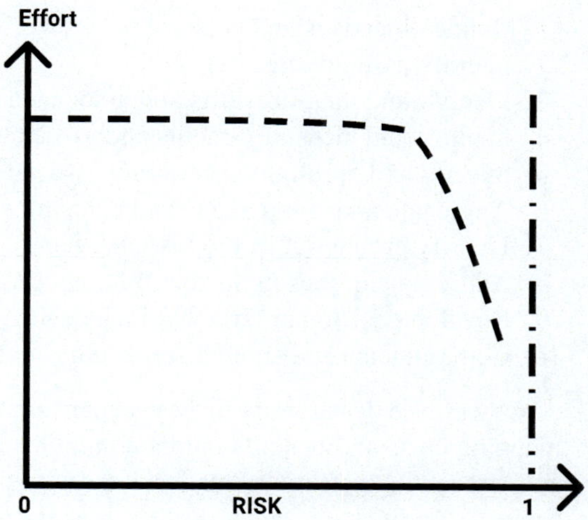

Fig. 5.1 Risk averse behavior (courtesy of ADA, Inc.)

overall public considered the same risk comparison differential to be substantially greater, 20%, 21%, 47%, and 72%, respectively.

Let us now formulate the quantitative aspects of cost-benefit analysis. The NPV of a VBD feature can be expressed as:

$$\mathrm{NPV} = \Sigma_{t=0}, T\, D_t \left\{\alpha(B,t)\left[B_t - \beta(C,t)C_t\right]\gamma(R,t)R_t\right\}, \tag{5.5}$$

where B, C, R = {benefit, cost, risk}; B_t, C_t, R_t = {benefit, cost, risk} at time t; $\alpha(B, t)$, $\beta(C, t)$, $\gamma(R, t)$ = factors, one for each, to convert the values of {benefit, cost, risk} onto a common scale so that their relative values can be added, subtracted, etc.; D_t = discount value of {benefit, cost, risk} adjusted for time t, i.e., the value of a benefit B1, may be less at a future time t' by a Facto r of $D_{t-t'}$; $D(t)$ 1/(1 + r) (t–t').

In applying cost-benefit analysis to VBD features, certain adaptations and elaborations are necessary. One of these is to account for multiple stakeholders. We enumerated an example set of stakeholders for Fallingwater in Chaps. 2 and 3. Thus, formula (5.5) needs to be adjusted to represent multiple costs–benefits–risks, each one for a different stakeholder (i, j, …):

$$\mathrm{NPV} = \sum t = 0, T \sum j \sum i\, Dj_t \left\{\alpha j(Bji,t)Bji_t - \beta j(Cji,t)Cji_t - \gamma j(Rji,t)Rji_t\right\} \tag{5.6}$$

When there are multiple dimensions of risk with constant costs and benefits, the difference between their respective NPV in (5.6) can be estimated through the following modified formulation:

$$\mathrm{NPV1} - \mathrm{NPV2} = \sum t = 0, T \left\{\Sigma j \Sigma i\, Dj_t\, \gamma j(Rji1,t)Rji_t - \Sigma j \Sigma i\, Dj_t\, \gamma j(Rji2,t)Rji2_t\right\} \tag{5.7}$$

In prescriptive terms, we recommend the following normative procedure for conducting cost-benefit analysis (Crouch and Wilson 1982, pp. 107):

1. Define objectives and goals.
2. Identify assumptions.
3. Identify and measure costs and associated risks.
4. Identify and measure benefits and associated benefits.
5. Specify and highlight uncertainty associated with risks.
6. Aggregate and compare cost and benefit, subject to associated risks.
7. Identify groups subject to loss and gains.
8. Address inequities in the distribution of costs and benefits.
9. Develop decision criteria that include any applicable constraints.
10. Communicate and summarize results.

Other obstacles in adapting cost-benefit analysis to support DAV analysis include mapping costs and benefits onto a common scale, quantifying ordinal evaluations, and conducting sensitivity analysis, three topics we will review in some detail in Chap. 8. Ordinal scales can be instrumental in finding common ground between unlike parameters to be measured. Say we are looking at benefits of Swiss Re Tower (Chap. 10) on Norman Foster's career versus the costs of constructing and maintaining the property. If we map both parameters onto a five-point ordinal scale of high, medium-high, medium, medium-low, and low and we assign a corresponding cardinal value to each of these points, say 0.99, 0.75, 0.50, 0.25, and 0.01, respectively, we can aggregate the two parametric dimensions of maintenance cost and benefit to career.

This is literally adding apples and oranges. Contrary to common belief, this is possible if we find a shared scale, as in the case of our other metaphors: calories, pesticide content, or nutrients. The weakness of our approach is that the five-point scale we use may create untenable sensitivities. What would happen if we assign 0.9 to high and 0.1 to low? Would the aggregated cost-benefit analysis value for these two dimensions be different? This is the topic of sensitivity analysis, to be discussed later in Chap. 8.

VBD Elicitation Methods

As we demonstrate for Swiss Re in Chap. 10, estimating the parameters of DAV analysis after the fact—costs, benefits, and risks associated with VBD features for all stakeholders at different levels of impact—is relatively straightforward. In this case, there is actual data to calculate the added value of design.

How about designs that still are on the drawing board or the planning stage? How can we estimate these parameters for DAV analysis? In fact, isn't it even more critical to have accurate estimates of these parameters at a time when the investment in their VBD features is still being considered? The decisions made during the early stages in fact make or break the viability of these designs.

In the next chapter, we will cover several survey and interview techniques that can be used to elicit and estimate VBD parameters not only after the fact but also before a design is finalized. The techniques we review include interview, ethnographic observation, cognitive walkthrough, brainstorming, and prototyping.

Exercise 4

The table below contains formulae that can be used to estimate the values for cost–benefit analysis, risk analysis, VBD analysis, NPV analysis, correction coefficients, and multiple dimensions of risk. Familiarize yourself with these formulations and select one to apply to data about a building delivery case found either in your own sources or ones provided in this text.

Formulas for cost–benefit and risk analysis calculations
1. The norm: $\{\alpha(B,t)B(t) - \beta(C,t)C(t)\}$; where B is benefit, C is cost, at time t, with α and β as coefficients of conversion into a common unit, for benefit and cost, respectively, for a given design feature and stakeholder
2. Value of a design decision-making strategy with risk: $\{\alpha(B,t)B(t) - \beta(C,t)C(t) - \gamma(R,t)R(t)\}$; where t is time; and α, β, and γ are coefficients of conversion into a common unit, for benefit, cost, and risk, respectively, for a given design feature and benefiting stakeholder
3. Objective of Value Based Design: $\max\{\Sigma_{k,l}[\Sigma_{j,i}(B_d - C_d)R_d]\}$; where B_d are the benefits subject to risk R of a decision d, C_d are the cost subject to the same risks, and R_d are the risk factor involved in a decision d, k and l define the range of costs and i and j the range of benefit involved with the decisions.
4. Objective of Value Based Design with correction coefficients: $\max\{\Sigma_j[\Sigma_l(\alpha B_d - \beta C_d)\phi R_d]\}$; where α, β, and Φ are correction coefficients to map B_d, C_d, and R_d into a common cardinal scale
5. The C-BA of the Net Present Value (NPV) of a Value Based Design feature: $\text{NPV} = \Sigma_{t=0},T\, D_t\{\alpha(B,t)[B_t - \beta(C,t)C_t]\gamma(R,t)R_t\}$; where B, C, R = {benefit, cost, risk}; B_t, C_t, R_t = {benefit, cost, risk} at time t; $\alpha(B, t)$, $\beta(C, t)$, $\gamma(R, t)$ = factors, one for each, to convert the values of {benefit, cost, risk} onto a common scale so that their relative values can be added, subtracted, etc.; D_t = discount value of {benefit, cost, risk} adjusted for time t, i.e., the value of a benefit $B1_t$ may be less at a future time t' by a Facto r of $D_{t-t'}$; $D(t)$ 1/(1 + r) (t–t')
6. The C-BA of the Net Present Value in Value Based Design feature representing multiple cost-benefits-risks for different stakeholders ($i, j, \ldots$): $\text{NPV} = \sum t = 0,T \sum j \sum i\, Dj_t\{\alpha j(Bji,t)Bji_t - \beta j(Cji,t)Cji_t - \gamma j(Rji,t)Rji_t\}$
7. Estimating the differences between multiple dimensions of risk with constant costs and benefits: $\text{NPV1} - \text{NPV2} = \sum t = 0,T\{\Sigma j\,\Sigma i\, Dj_t\,\gamma j(Rji1,t)Rji_t - \Sigma j\,\Sigma i\, Dj_t\,\gamma j(Rji2,t)Rji2_t\}$

Chapter 6
Elicitation Methods

The first manifestation of a VBD feature in a project arises in the earliest stages of the design process when the idea for a design is being conceived. This is the design requirement specification, or architectural programming, stage when the users' needs, wants, and potential design features are considered. Design needs are the true expression of what is necessary in a new design. User needs, whether functional, aesthetic, or economic in nature, are the true expression of what is necessary in a design and what are essential to the user or client. The simple question "what do you need?" is the prototypical query to identify design requirements. However, the answer to this simple query is often insufficient since clients and users do not have a full understanding of what is needed. Since the design process itself may illuminate overlooked requirements, later on, the preliminary list of requirements may need augmentation.

Deeper background research is often required to understand the root-causes of initially expressed needs. And in the process of such investigations, specific VBD features may rise to the surface and be added to the wish list. The elicitation methods for uncovering VBD features include structured interview, ethnographic observation, cognitive walkthrough, brainstorming, and prototyping (Table 6.1).

Clients' need, rarely, if ever, require VBD features. Their charge is to express those features that are fundamental to the intent of the design. Wants of the client or the user, on the other hand, can include aspects of the architectural program that go beyond the basics and hint at VBD features. Sophisticated users, in particular, can often state uses of the design that are not necessarily readily available. For example, wishing for a museum that will attract patrons due in part to its architectural features, as in the case of the Guggenheim in New York, by Wright, or Guggenheim in Bilbao, for that matter, by F. O. Gehry, is a good example of this category. Structured interview is the most suitable elicitation method for this case.

Potentially suitable design features of a project that are suggested by the designer(s) at an early stage of the design process can often result in VBD features. Swiss Re, in addition to its great financial success, has attained great acclaim as a capstone building in London. Cognitive walkthrough, brainstorming, or prototyping

Ö. Akın, *Design Added Value*, https://doi.org/10.1007/978-3-030-28860-0_6

Table 6.1 Requirement Specification Procedures for VBD and Elicitation Surveys

Procedural step	Description	Survey type
Assess the overall problem	Describe the nature of the design needed and its requirements	Cognitive walkthrough
Establish customer or user profile	Determine personal, intellectual, professional, profiles	Structured interview
Understand the user environment – i.e., limits	Describe the manner in which users will engage the new design	Ethnographic observation
Recap to client/user for shared understanding	Develop report and share	Cognitive walkthrough
Assess all design feature possibilities	Enumerate the design features with no regard for VBD features	Brainstorming
Assess only VBD feature possibilities	Tease out the VBD features envisioned by the requirements	Prototyping
Resources needed to realize the VBD	Assess performance reliability, and support needed for VBD	Cognitive walkthrough
Other requirements, i.e., risk management for VBD	What risks are possible and what are their expected consequences	Brainstorming
analyst's summary	Report to summarize all of the above	Reporting

techniques serve best for facilitating requirements elicitation for such purposes where features beyond functional requirements need to be understood (Table 6.1).

A more complete compendium of requirement types, some of which are included in Table 6.1, can be adopted from design requirement specification literature. Information elicitation usually takes nine steps before it is fully completed: (1) assess overall problem, (2) establish customer profile, (3) understand the user environment, (4) develop shared understanding of needs, (5) assess design features, (6) assess VBD features, (7) enumerate required VBD resources, (8) consider VBD risk management, and (9) provide a summary of this analysis.

All survey techniques that involve stakeholders and subjects, human or otherwise, are subject to review and approval of their methods to ensure that no harm is done to their participants.

Structured Interview

A structured interview is a questionnaire conducted face-to-face. It is not as cut and dried as a formal questionnaire you may receive in the mail or through email, but it is a structured set of questions asked in a given order. It has the advantage of allowing the interviewer and the interviewee to pursue important issues that may come up during the course of the interview. A structured interview provides the best means for getting right to the crux of the "need" issue in a design problem. Requirement specifications can be directly gathered from the users, clients, or others being interviewed. In addition, some aspects of the design problem not anticipated by the initial question set can be posed as add-on questions at an appropriate point during the course of the face-to-face interaction.

Design of structured interviews follows a general procedure including sample selection, background research, design of questions, solution possibilities, invitation of participants, scheduling of interview, written instructions, demographic information gathering, and finally the administration of the interview. Sampling is a field in its own right and standard texts in this area should be consulted to make sure that the sample selected for the interview is representative of the population it is expected to represent.

Researching the background of information relevant to the topic of investigation can help formulate substantive and relevant interview questions. Potential answers should also be considered in order to plan for digressions and sidelines that may come up. Interviewees must be contacted in advance and the intent of the interview must be explained while obtaining their consent to participate. Prior to the commencement of the interview, participants must be briefed on the intent, consent forms, and the format of the interview. During the implementation phase, interviewers must refrain from interjecting their own opinions and feel free to go with the flow of the "conversational" mode in the event of useful digressions.

While a structured interview is best suited to the specification of the design problem and establishing the customer profile, it can also accommodate the aspects of DAV analysis. In fact, it can be a reliable fallback for many of the other steps of VBD feature elicitation, including assessing the overall problem, understand the user environment, assessing design features, assessing resources needed to realize the VBD, and risk management.

A well-prepared structured interview can be thorough, covering all of the desired aspects of the issues being explored, without overpowering the participants. This is because of the latitude allowed for digression. However, care must be taken to make any interview as short as possible, certainly no longer than 1.5 hours. If necessary, the interview can be broken up into shorter parts administered separately, at different times, with different participants or through different elicitation methods.

Conducting interviews with multiple participants on the same topic is subject to the "law" of information saturation. As subsequent participants are interviewed, redundant information will most certainly come up. In other words, there will be shared views among participants on similar topics. If we chart this in a graph with the abscissa as "number of observations" or interviews, and the ordinate as the acquired "information content," the content of additional information for each new observation will diminish until it asymptotes at a certain level (the dashed line in Fig. 6.1).

The questions included in a structured interview should strike a balance between relying on past knowledge about the topics being investigated and unbiased inquiry. We often have some idea about what sorts of answers we would normally get, which represents a bias that may or may not be confirmed by the facts of the situation. Therefore, care must be taken to frame questions in a way that does not condition the likelihood of any particular answer. Depending on the responses, if unintended bias should emerge, the questions must be reframed to eliminate this bias.

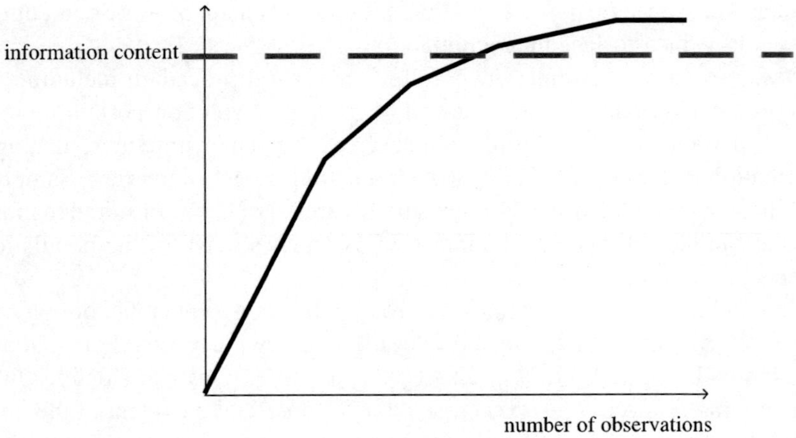

Fig. 6.1 The "law" of elicited information accumulation

This is one of the reasons why simply sending out the questionnaire is not a good alternative to a structured interview following the questionnaire for the purposes of requirement elicitation in most cases. In a remotely administered questionnaire, there is no good way of adapting the query to the context of the participant. Furthermore, the response rate to questionnaires is generally poor, around 10–20% of the total number of solicitations. This is confirmed in a variety of domains when the participants are members of the same population.

Brainstorming

Brainstorming is the perfect technique to discover that which, at first, does not meet the eye. Among the steps of the VBD elicitation process the two that are most suitable for applying brainstorming are "assessing the features of the design" and "managing VBD risks" steps.

Brainstorming is best conducted with adherence to a handful of golden rules: generate as many ideas as possible, allow imaginations to soar, permit no criticism, or debate while generating ideas, and mutate and combine ideas by piggybacking on previously stated ones. To initiate a brainstorming session the first step is to state the objectives of the session to the participants. Next, the participants are urged to keep it informal and use their own words. A scribe keeps accurate notes of keywords that represent ideas generated. Finally, session is to be kept no more than one or one and a half hour, certainly not more than two hours.

Once ideas are generated in the brainstorming session, they have to be edited and organized into a coherent format. The open, unabbreviated format of these sessions is conducive to generating both inventive and unsuitable ideas. That is precisely the intent: by allowing all kinds of ideas to emerge without restriction, it is possible to harvest inventive ideas that otherwise may not be suggested at all. Therefore, the post-session challenge is to separate the wheat from the chaff.

First, the list of keywords and clusters of keywords associated through idea-piggybacking are pruned into a shorter more relevant list. Here judgment is exercised to separate ideas into three categories: those that would not work, those that would, and those ideas that are exceptional without regard for feasibility. Authorship of ideas if present in the notes should be eliminated in this stage to avoid unintentional biases from influencing judgment.

Once the pruning process is complete, the keywords and phrases should be sorted into two categories: (1) technology aspect—new architectural features, enhanced existing features, occupancy issues, and any related aspects; and (2) domain aspect—customer/user issues, budget, schedule, context, and any other related concerns. During this stage, further elaboration of the key ideas and features is encouraged and supported. Greater detail and specificity should be provided.

Finally, these ideas have to meet the filter of reality. It is possible that more ideas than those which can be realized may be generated. One of the best ways of reducing scope of this kind is to prioritize the keywords and ideas. This is potentially the most contentious aspect of post-session processes. Participants should be issued limited resources, say $100, to spend on their preferred keywords or ideas. This sort of "voting" for your preferences should be done only once. Multiple cycles tend to lead to collusion and coercion behavior as a function of the previous cycles and do not seem to yield good results. Allowing participants to allocate their budget as they like, all on one idea versus spreading on multiple ideas, often help them to prioritize in different ways.

Even with a well-run prioritization session there are challenges to be overcome. Ordinal qualifications of elicited items do not provide a clear metric to decide which entities can be eliminated. Concepts denoting significance, such as "critical," "important," and "useful," can be used to further discriminate the data. Alternatively, hierarchies can be constructed to nest certain concepts inside others in order to simplify the synopsis of findings. Creating hierarchic priority schema is a vast area of research in and of itself such as the Analytical Hierarchy Method (AHM) (Saaty 1980).[1] This approach is based on pairwise comparison of options by predefined and prioritized criteria. Prioritization is accomplished through the repeated pairwise comparison.

Based on a brainstorming session we conducted on the facilities management programs of a major university, we developed major service options like providing support 7–24 service to the occupants in the university. Several alternative approaches emerged including using human operators, versus automating the entire system on a digital platform, and cooking up a hybrid system.

Several criteria or evaluation emerged: 100%accessibility, error-free processing, friendly interaction, robust operations (nearly no failures), and ease of dovetailing with existing 9 to 5 operations. We created subgoals to assess robustness rating for

[1] Saaty, T. The Analytic Hierarchy Process: Planning, Priority Setting, Resource Allocation, New York, NY: McGraw-Hill, 1980.

Table 6.2 Estimating weights for subgoals using AHM (Saaty 1980)

Robustness rating	Human	Auto w/oversight	Human w/Toggle	Auto w/backup	Hybrid
Human	1	2	3	4	1/2
Auto w/oversight	1/2	1	1/5	1	1/2
Human w/toggle	1/3	5	1	3	3
Auto w/backup	1/4	1	1/3	1	1/2
Hybrid	2	2	1/3	2	1
Σx	4.08	11	4.87	11	5.5

Table 6.3 Hierarchical ordering of subgoals using AHM (Saaty 1980)

Robustness rating	Human	Auto w/oversight	Human w/Toggle	Auto w/backup	Hybrid
Human	0.081	0.455	0.205	0.273	0.545
Auto w/ oversight	0.490	0.182	0.068	0.182	0.182
Human w/ toggle	0.245	0.182	0.616	0.364	0.091
Auto w/ backup	0.123	0.091	0.041	0.091	0.091
Hybrid	0.061	0.091	0.068	0.091	0.091
Σx	0.081	0.455	0.205	0.273	0.545

the above automation options using AHM. These included (1) human, (2) auto w/ human oversight, (3) human w/auto-pilot toggle, (4) auto w/human backup, and (5) hybrid. First, we estimated weights through AHM to assess the relative importance of these subgoals (Table 6.2) and then calculated ordered coefficients for each (Table 6.3).

Brainstorming is one of the most democratic ways of culling ideas from a large group of constituents and then organizing them into a workable format, including idea convergence, comprehensive written records, incorporating a wide bandwidth of ideas and potential solutions, and to accommodate out-of-the-box thinking.

Prototyping

Prototyping is an approach that is best suited for assessing VBD feature possibilities in a design project. It is useful in articulating prototypical design requirements for facilities. It involves the development of an advanced model or representation of the product indicating what it would look like or perform as. It is also helpful in correcting known design pitfalls by getting them out in the open, such as "yes-but," "undiscovered ruin," and designer bias issues.

Yes-but responses by users imply indecision about the priorities associated with design features. Prototypes can help participants decide once and for all. Undiscovered ruins are emergent features that are hard to identify until a simulation of the design is at hand and users interact with them. And design biases are designer's blocks that can be impediments to thinking outside the box. Prototypes can be instrumental in exposing these biases and overcoming them.

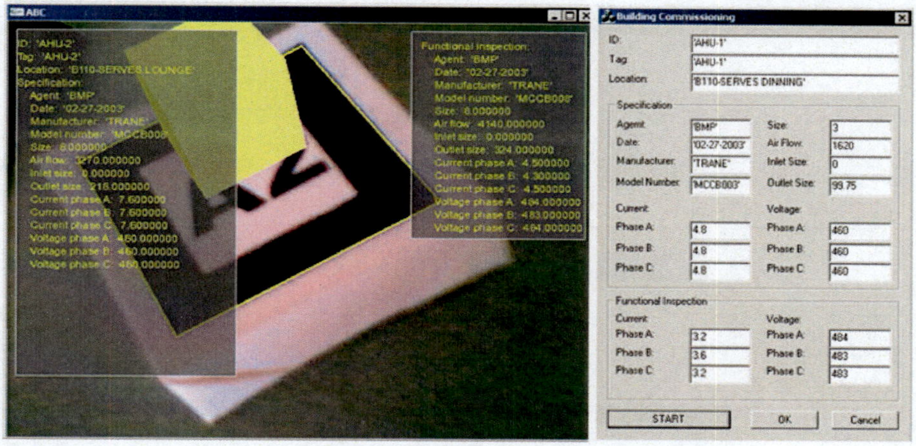

Fig. 6.2 Prototype—RFID-based AR—Visualization for Field Data and Training Lee, Sang Hoon and Ömer Akın (2010)

In a project where we envisioned virtual reality and augmented reality as vehicles for improving building operations and maintenance problems, we developed a prototype that simulates what would happen if indeed we had such a tool. This tool would have a visual display on a digital screen that juxtaposed geometric shapes, interiors of buildings. When a tag appears on a visual device, such as the tag "A2" in Fig. 6.2, the user can enter the virtual space to test their ideas about "yes-but," "undiscovered ruins," and "designer bias" categories. In addition, it would display textual, static information from databases and dynamic information from live feeds off of sensors on the same display. To assist operations, we also developed a storyboard type illustration of the information that could be obtained from and entered into such a system (Fig. 6.2 left half).

Prototypes come in several types (Davis 1995b). Some are focused on the user interface aspects of a design, such as the digital tools we just discussed. Others may be more interested in some performance issue like minimizing energy use in a light fixture. Here the prototype would illustrate how energy would be saved rather than how the fixture would look. Some prototypes are evolutionary, meaning that as work continues on the project the prototype is enhanced and further developed to create versions closer and closer to the final design or product. Alternatively, a prototype may be designed to be a throwaway; in other words, after it is used for the limited purpose it is designed, it is discarded.

Prototypes can be vertical and horizontal. Vertical prototypes dig deep into all aspects of a limited section of a large project, say the lobby of a hotel design, or the clean rooms in a lab building. Alternatively, horizontal prototypes focus on a broad coverage of all major parts of a project, bringing out the interrelationships between these major parts rather than the minutia of detail each one of them may entail, for example, focusing on circulation patterns of the hotel design.

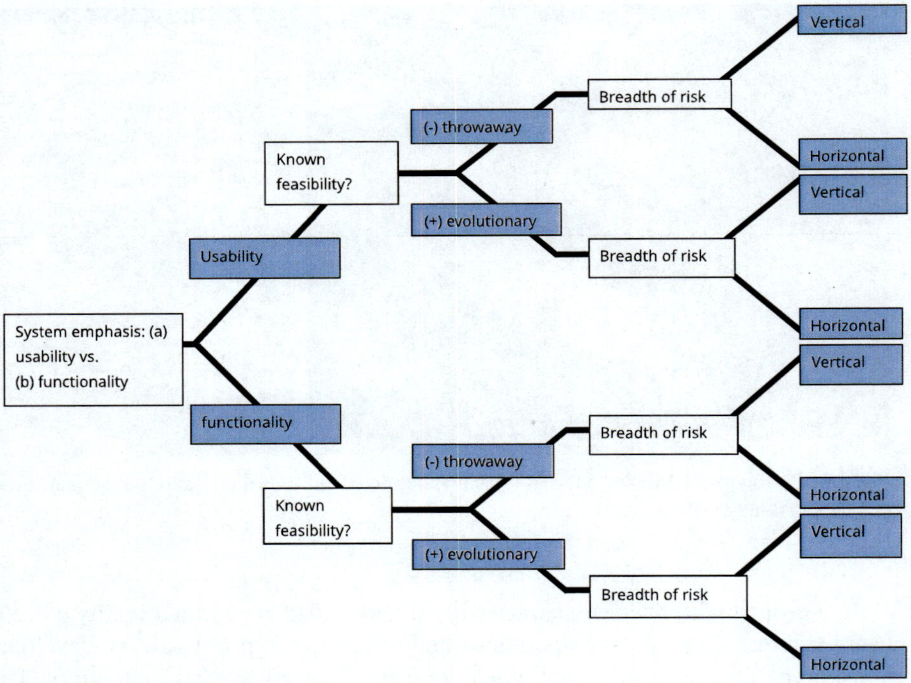

Fig. 6.3 Prototyping decision tree

Figure 6.3 shows a decision tree for prototype designers to select the right type they should build by going through three tiers of decision. Are we interested in usability or functionality or performance issues? Are we interested in a prototype that can be evolved into the final product or one that we can throw away, once we gain the benefit it provided? And are we interested in a narrow slice of the project or a breadth of higher-level issues?

As an elicitation tool prototyping can be used to explore the usability emphasis and test the availability of tools for occupant comfort or as gateways to functional requirements. Assumptions by the designer can be built into a prototype as a catalyst for elicitation of requirements from the client. These assumptions can be used by the client to communicate design issues with the designer. Prototyping process can transform unknown design features or requirements into fuzzy ones and the fuzzy ones into well-known and well-understood ones.

Cognitive Walkthrough

Cognitive walkthrough can be used to assess the overall problem, recap the problem to client/user for better shared understanding, and to assess the resources needed to realize VBD features. Cognitive walkthroughs are done by domain experts, individuals who know as much about the building type including the design problem as

Table 6.4 A Cognitive Walkthrough Session to Develop a Prototype System for SEED (Courtesy of ADA, Inc.)

Participant-A:	"Let's create a specification unit (SU) for the lobby"
ACTION:	pop up menu --> create spec unit, type name 'lobby'...
Participant-B:	"Can you just copy them instead of typing (it) each time...?"
Participant-A:	"Yes, it is another alternative, but when you copy and paste the counter will not change. Then you have to change (update) the counter for each copy individually."
NOTE:	this is an apparent usability flaw. ......
Participant-C:	"So when you create a new SU, it does not automatically pop up a new window..."
Participant-A:	"Yes, it does, do you want to see it..."
Participant-C:	"I am not asking when you edit, I mean when you first create one."
NOTE:	This highlights state specific conditions ... ...

anyone. They are invited to examine a legacy design or a prototype developed for the problem at hand.

In a cognitive walkthrough session is planned which is carried out by these domain experts. In this session, a predefined script is followed to assess the system or design under investigation. The purpose of the session is to develop a specific, feature by feature, evaluation. The session must be facilitated carefully by someone experienced in running workshops. The information generated during the workshop must be recorded and later analyzed for evidence to support evaluative inferences.

In developing a prototype system for the early design of a system called Software Environment for Early Phases of Building Design (SEED) we conducted a cognitive walkthrough. Table 6.4 demonstrates a recorded discussion segment from this session.

In the span of a couple minutes through this walkthrough we see two instances of evidence—a "usability flaw," and "state specific conditions"—pointing to the characteristics of the system under assessment. Cognitive walkthroughs are extremely effective in developing a rapid understanding of design features prior to the commitment of large quantities of time, funds, and other resources.

Ethnographic Observation

Ethnographic observation is probably the most elaborate and resource intense of all elicitation methods. They are most suitable for understanding the user environment in its entirety and its limits. Ethnographic observation is a structured and systematic way of codifying user behavior as it takes place in its natural context. It uses the actual events and interactions in a facility rather than creating simulations or prototypes which are mere abstractions of the real thing. In addition to being naturalistic, ethnographic observations aspire to be unobtrusive. They try not to interfere with the behaviors they codify. It must be noted for the record that this is very difficult to guarantee, since any observation by definition impacts that which is being observed and the personal opinions and biases of the observer may influence the data collection despite all preventative measures. Anchoring all the findings in the factually collected data is therefore essential in conducting successful ethnographic observations.

To achieve these goals, ethnographic observation techniques involve extensive data recording either through camcorders or other recording devices. But data collection is not limited to such recordings as it is a multi-granulated approach including a variety of sources of data: graphical recordings, physical products and tools used by participants, written records, databases, and traces of human behavior.

Flemming et al. (1997) conducted an ethnographic study of user behavior in using a CAD system called Microstation by Bentley Systems, Inc. The researchers observed how designers use this system to draft architectural drawings. They did video recordings of multiple draftsmen and supplemented this with records of hard-copy drawings, project information, and other data. They found that draftsmen used, what they call, a "T-square Metaphor." Some of the data that they found which supports this conclusion is blow by blow creation of a drawing of fire protection encasing of a column in plan-view (Fig. 6.4) as if they were using a T-square to draw rather than take into consideration the shortcuts available in most automated drafting systems. The sequence of lines resembles what one would do if one was working with ink and t-square on velum: draw parallel lines first going from top to bottom and left to right, make sure ends of lines meet, and repeat the same for the next object. Whereas, in CAD lines can be drawn in any order, trimming is not necessary, copying and pasting is all it takes to create the next object of the same kind.

In another ethnographic observation study, we shadowed the operations and management (O&M) personnel of a major university (Fig. 6.5). First, we arranged with the field workers to shadow them. Next, we audio recorded about 50 work orders being carried out in the field. Next, we transcribed our data and categorized it into trades, types, durations, and so on. Next, we developed a model process that consists of four components: *pre-dispatch, reconnaissance-intervention, fix*, and *post-dispatch*. Finally, this model was compared and calibrated alternative, information-based, process applications.

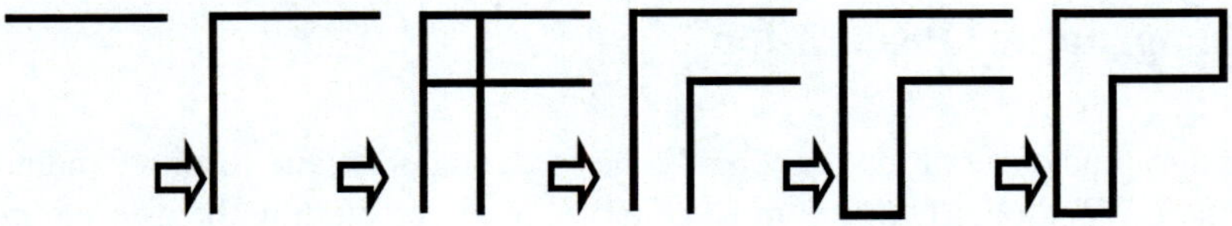

Fig. 6.4 Microsoft-based drafting sequence of fire protection encasement

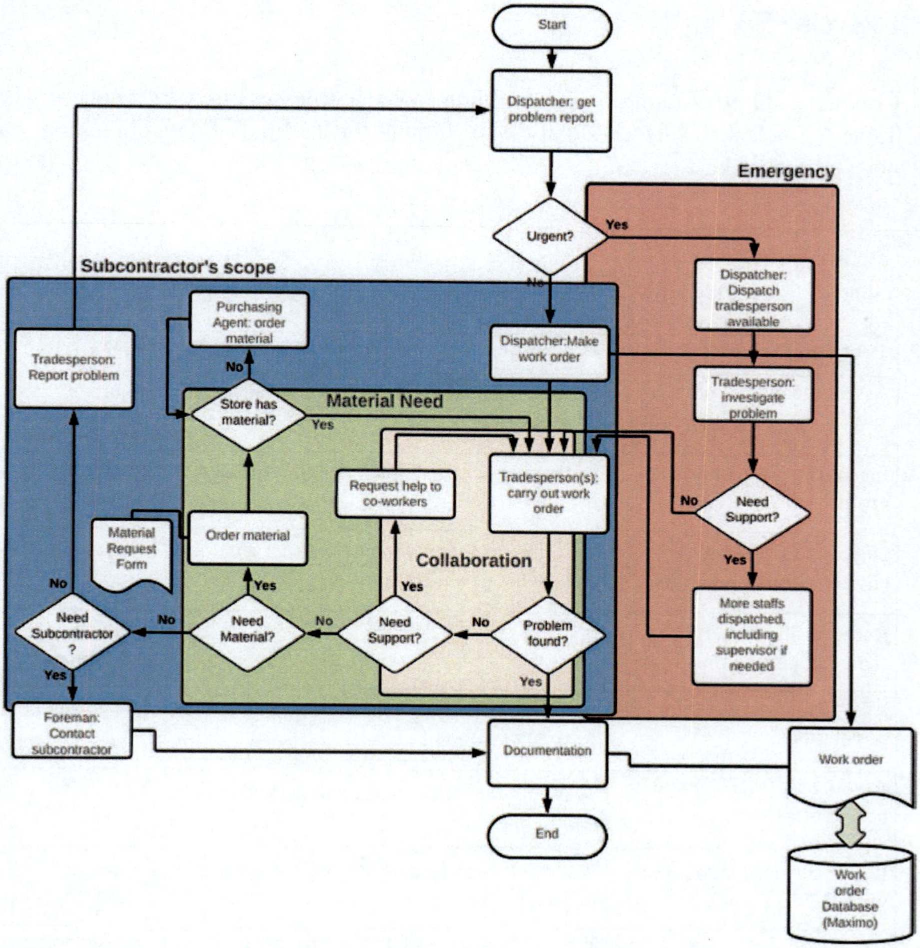

Fig. 6.5 The process model for operations and maintenance based on the shadowing tasks (Courtesy of ADA, Inc.)

Exercise 5.1

Conduct data elicitation about a building in order to assess its CPD process. Use Table 6.5 below to select a survey type suitable for the kind of data elicitation you need to undertake.

Table 6.5 Survey types for data elicitation in the capital project delivery process

Procedural step	Description	Survey type
Assess the overall problem	Describe the nature of the design needed and its requirements	Cognitive walkthrough
Establish customer or user profile	Determine personal, intellectual, professional, profiles	Structured interview
Understand the user environment – i.e., limits	Describe the manner in which users will engage the new design	Ethnographic observation
Recap to client/user for shared understanding	Develop report and share	Cognitive walkthrough
Assess all design feature possibilities	Enumerate the design features with no regard for VBD features	Brainstorming
Assess only VBD feature possibilities	Tease out the VBD feaures envisioned by the requirements	Prototyping
Resources needed to realize the VBD	Assess performance reliability, and support needed for VBD	Cognitive walkthrough
Other requirements, i.e., risk management for VBD	What risks are possible and what are their expected consequences	Brainstorming
Analyst's summary	Report to summarize all of above	Reporting

Exercise 5.2

Use Table 6.6 below to determine the source(s) from which you can best acquire the data you are seeking to assess its CPD process. Take into consideration the objectivity of the source, your ability to access the source, and issues of privacy and confidentiality with which you need to treat the data you obtain.

Table 6.6 Sources for data acquisition in the capital project delivery process

Building lifecycle stages	Design professionals involved	Other participants involved
Architectural programming	Architects, facility programmers, planners, appraisers, specialty designers	Clients, users, owners, regulators
Preliminary design	Architects, civil engineers, mechanical engineers, planners, cost analysts, specialty designers	Clients, users, owners
Design development	Architects, civil engineers, mechanical engineers, specialty designers	Clients, users, owners, regulators
Construction documents	Architects, civil engineers, mechanical engineers, contractors, data management specialists, specialty designers	Regulators
Construction	Contractors, sub-contractors, specialty manufacturers, architects, civil engineers, mechanical engineers, data specialists	Regulators, local officials, owners
Commissioning	Commissioning authority, contractors, mechanical engineers, architects, operation and maintenance staff	Users, clients, owners
Occupancy	Contractors, space planners, interior designers	Users, clients, owners
Operations and maintenance	Contractors, operations and maintenance staff	Users, clients, owners
Continuous commissioning	Commissioning authority, mechanical engineers, architects, operation and maintenance staff	Users, clients, owners
Retrofitting	Architects, facility programmers, planners, contractors, appraisers	Clients, users, owners, regulators
Re-occupancy	Space planners, interior designers	Users, clients, owners
Decommissioning	Contractors, planners, appraisers, developers	Users, clients, owners

Chapter 7
Pre facto and Post facto Analysis

Pre facto Analysis: (De-)value Engineering

The classical *value engineering* methods provide a suitable benchmark for function-feature-cost analysis in VBD. In order to fully unravel this relationship between what at first might be construed as "strange bed fellows," if not outright antagonists, we need to take a closer look at value engineering in its own terms.

> *Value Engineering is the systematic application of recognized techniques which identify the function of a product or service, establish a monetary value for that function and provide the necessary reliability for it at the lowest overall cost.*[1]

Value engineering deals with two central concepts *function* and *cost*. Value engineering aims to minimize cost without sacrificing function. Functionality, whether it is indispensable or incidental to the core purposes of a facility, is the primary objective of any facility or its design features in value engineering. Hence, in the design, construction, and the occupancy stage the functions of the facility are analyzed in detail. Great deal of attention is paid to distinguishing *primary* functions from *secondary* ones. Next, the costs of the design components which correspond to these functions are estimated (Fig. 7.1).

In VBD, there is a third category of critical ingredients: *design features*. What is tacit in the value engineering model, meaning design components, is explicit in VBD. Here the task is to perform the same analysis as in value engineering and then estimate the value adding potential of novel design features or components. VBD recognizes that some design features are eliminated through value engineering because they are not proven to be of sufficient benefit to the project. However, these features may in fact add to the value of the project in unfamiliar and unexpected ways because they are novel and their use in the field is not properly understood yet.

[1] Mudge A (1989) Value engineering. J. Pohl Associates, New York, NY.

Ö. Akın, *Design Added Value*, https://doi.org/10.1007/978-3-030-28860-0_7

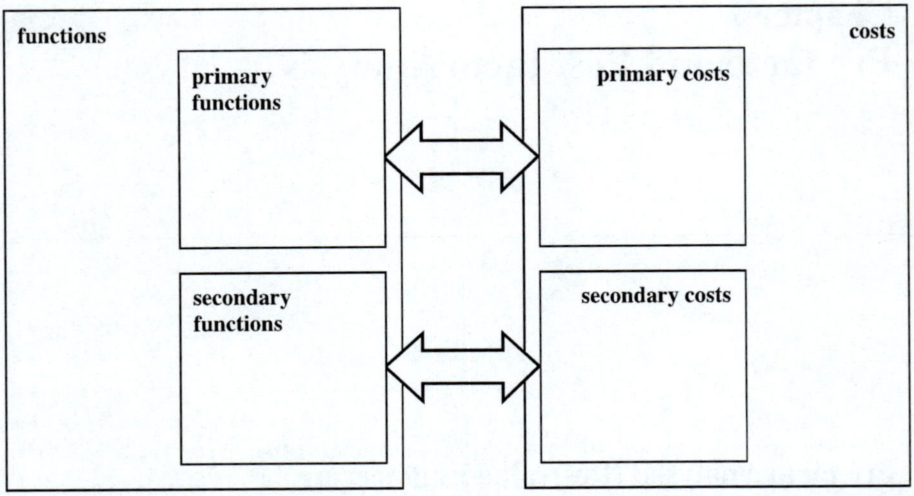

Fig. 7.1 Two-part value engineering model (Mudge 1989)

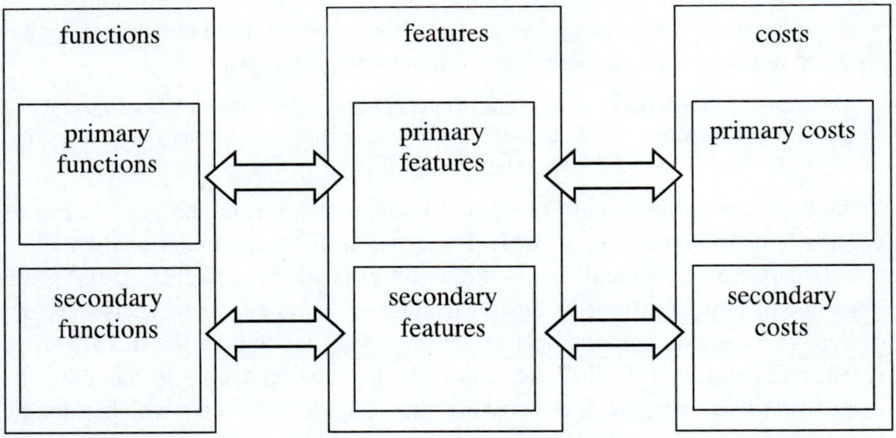

Fig. 7.2 The three-part value-based design model

Arguably, such value is emergent in many creative design proposals, and it is justifiable that the two-part model of value engineering needs to be extended to the three-part model of VBD as shown in Fig. 7.2.

The VBD model is not only more comprehensive in its scope than value engineering, but it is also more accurate in estimating costs. Functions are abstract entities that cannot be directly valued in monetary terms. It is either through the utility they provide or the physical components that have to be created to make the function possible. The latter method is more accessible at least in the field of AEC through the commonly available methods of material take-offs and cost estimation. In this model, design features serve the critical role of linking functions to costs through physical components that can be enumerated, measured, and estimated (Table 7.1).

Table 7.1 Comparing value engineering with VBD

Value Engineering	Value-based Design
Conception phase	
Value resides in the flexible functions	Value resides in the flexible features
Value resides in the selling functions	Value resides in the selling features
Isolate and minimize costs of flexible and	Include and maximize flexible and
Look for *selling features*[xliii] that can be	Look for the value of selling designs
Look for flexible functions that can be	Look for the solution values of
Evaluation phase	
Select complete functional sets	Select corresponding design sets
Add costs for each set [RCB]	Add costs for each set [RCB]
Disadvantages [breakdown by	Advantages [benefits, lifecycle
Associate costs and benefits	Consider Value Based Design

What Is Function?

Let us now consider the definition of *functions* in value engineering. There are three categories of definition each with a pair of concepts: composition of (1) functional descriptions, (2) functional types, and (3) significance levels of functions. Functional descriptions are composed of two natural language word types: verb and noun. "Support the roof," "provide comfort," "save energy," and "maintain privacy" are some examples. In each case the first word is a verb and the second a noun.

Some of these verbs are *active* or strong verbs, like "support" and "save," while others are soft and *passive* like "provide" and "maintain." Similarly, the nouns can also be classified. "Roof" and "energy" are *concrete* and *measurable* while "privacy" and "comfort" are *abstract* and *immeasurable*. These nouns when paired up with the verb categories in a way that action verbs and measurable nouns are grouped, we get *work* functions. *Sell* functions, on the other hand, are expressed with passive verbs and immeasurable nouns. Combinations of immeasurable nouns with action verbs and measurable nouns with passive verbs usually signal compound functional descriptions that should be broken down into more specific components to achieve a clean categorization into work and sell functions.

Finally, we can sort functions into secondary (or *support*) and primary (or *basic*) functions (Mudge 1989). This categorization is based on the significance of the function in meeting the requirements of a facility or its physical components and features. Basic functions include those that signify the primary purpose of a facility: a school to accommodate learning and teaching, a library to store and protect books, a factory to make industrial products.

Since each one of these facilities contain things that are essential to their main purpose like books, children, and merchandise, which need protection from weather, they have additional basic functions like structure, enclosure, conditioned air. Support functions like perfect heat and humidity conditions in the school or pleasing wall murals in the factory are nontrivial important requirements, but they are considered dispensable to varying degrees. Common belief dictates that basic functions can still be carried out without support functions. Note that a support function in one case, say humidity and temperature control, can be a basic function in another case, say in the archival collection room in the library. In the case of DAV analysis, work and basic functions are not likely to lead to features with added value while the sell and secondary functions usually do. Having indoor plumbing at home is not necessarily an unusual asset (in the twentieth century on) but an atrium space entry would be.

Functional descriptions are usually explicit during the requirement specification, or architectural programming, phase of design delivery (Table 7.2). Therefore, a functional analysis is most feasible during the early stages of design. In the later stages, the focus of design delivery is on components and costs. This is when the functional descriptions are mapped into costs *vis a vis* the components, and is the reason for emphasizing the mapping into components (or features) during DAV analysis.

What Is Cost?

Cost of facilities is a multifaceted entity (Table 7.3). The total cost of a facility is a combination of its first cost and use cost. The latter can be broken down into two costs: maintenance costs and operating costs, while the former into eight independent cost categories: (1) profit, (2) selling, (3) general administrative, (4) field service and miscellaneous, (5) indirect labor, (6) indirect material, (7) direct labor, and (8) direct material. Working from left to right in Table 7.3, one can now re-bundle these costs into other cost categories, like field service and miscellaneous, indirect labor, and indirect material costs constituting overhead costs; direct labor, and direct material costs constituting the prime costs, and so on.

It is often not possible to find the cost information for each of the low grain categories, but value engineering or VBD analysis can be conducted even with the higher grain categories, first cost, or use cost, so long as they can be broken down into physical components or features that constitute the innovative elements of the facility.

Table 7.2 Capital project delivery process and its participants

Building life cycle phases	Design professionals	Other participants
Architectural programming	Architects, facility programmers, planners, appraisers, specialty designers	Clients, users, owners, regulators
Preliminary design	Architects, civil engineers, mechanical engineers, planners, cost analysts, specialty designers	Clients, users, owners
Design development	Architects, civil engineers, mechanical engineers, specialty designers	Clients, users, owners, regulators
Construction documents	Architects, civil engineers, mechanical engineers, contractors, data management specialists, specialty designers	Regulators
Construction	Contractors, subcontractors, specialty manufacturers, architects, civil engineers, mechanical engineers, data specialists	Regulators, local officials, owners
Commissioning	Commissioning authority, contractors, mechanical engineers, architects, operation and maintenance staff	Users, clients, owners
Occupancy	Contractors, space planners, interior designers	Users, clients, owners
Operations and maintenance	Contractors, operations and maintenance staff	Users, clients, owners
Continuous commissioning	Commissioning authority, mechanical engineers, architects, operation and maintenance staff	Users, clients, owners
Retrofitting	Architects, facility programmers, planners, contractors, appraisers	Clients, users, owners, regulators
Re-occupancy	Space planners, interior designers	Users, clients, owners
Decommissioning	Contractors, planners, appraisers, developers	Users, clients, owners

A Summary Approach to Value Engineering in DAV

Value engineering provides a clear motivation for VBD, since owners and professionals alike prefer to take out the fat from their projects that can cause cost inflation. In DAV analysis, design features provide the basis for costing functions included in the design. Since features are not accounted for in classical value engineering, this fills the void that is difficult to fill otherwise. These features help define branding, and add value through features with no apparent exchange value or use in the functional realm of a building. As a result, these design features elevate value beyond the initially expected levels.

The conditions necessary for such value to be recognized include renowned architect, unconventional design, strong will to complete the project (sometimes against odds), and a resourceful patron that often comes with no budget limitations. In the end there is often a good dose of "spin" and "buzz" that results in strong branding of the building and all stakeholders associated with it (Toker 2003). In

Table 7.3 Cost categories and their mapping in facilities

<table>
<tr><td rowspan="10">Total Cost
[Life- Time Cost]</td><td rowspan="8">First Cost
[Selling Price]</td><td colspan="5">Profit</td></tr>
<tr><td colspan="4">Selling Cost</td><td rowspan="7">Cost of Goods</td></tr>
<tr><td colspan="4">General Administration Costs</td></tr>
<tr><td>Field service & Miscellaneous</td><td rowspan="3">Overhead Costs</td><td rowspan="4">Conversion Costs</td><td rowspan="5">Cost of Manufacturing Goods</td></tr>
<tr><td>Indirect Labor</td></tr>
<tr><td>Indirect Material</td></tr>
<tr><td>Direct Labor</td><td rowspan="2">Prime Costs</td></tr>
<tr><td>Direct Material</td><td>Material Costs</td></tr>
<tr><td rowspan="2">Use Cost</td><td colspan="5">Maintenance Costs</td></tr>
<tr><td colspan="5">Operating Costs – shipping, receiving, installation, etc.</td></tr>
</table>

order to assess the value of branding, we need to estimate not only the expenditures and gains, but also the confidence we can assign to these values in terms of risk (probabilities of accuracy), covered under pre facto analysis.

Pre Facto Analysis: Carry-Over of Estimates from Precedents

The approach of carrying estimates from a precented project is applicable if precedents can be found that match the VBD project at hand so closely that their estimated values can be applied with proper recalibration to the project at hand. This method requires that the precedents available provide the relevant data missing in the new DAV formulation. For example, if we were asked to do DAV analysis for a residential building with unusual, potentially trend setting features, designed by a notable architect, for a wealthy and ambitious client, and all data needed was not available either due to missing archival data or the fact that the project has not yet been undertaken, then we could reasonably consider Fallingwater by Wright, the Farnsworth House by Mies Van der Rohe, or the Villa Savoye by Le Corbusier as potential precedents that can assist our benchmarking.

The dimensions along which a VBD project must resemble the precedents include project size, scope, client, site, architect, costs, schedules, and facility program. At a minimum, the precedent must be well documented in terms of design features, costs of realizing these features, and benefits realized because of these features. Judged qualitatively, the similarity between them must be reasonably close—approximately within 50%.

When these conditions are met, the following procedure can be applied.

- *Features*—list the missing features, match features with corresponding functions distinguishing VBD from non-VBD functions, such as *work* versus *sell* functions and *basic* versus *secondary* functions.
- *Stakeholders*—list the missing stakeholders that have been affected by the features of this facility—at a minimum, include six levels: design professional, user, city or region, culture-media industry, commercial line, and political appeal.
- *Costs at Three Levels of Impact*—estimate the missing risks and costs at the immediate (first 2 years), medium (first 5–10 years), and long term (lifetime of building use) impact levels, as delta (differential) obtained by comparing against a benchmark.
- *Benefit at three Levels of Impact*—estimate the missing costs at the immediate (first 2 years), medium (first 5–10 years), and long term (lifetime of building use) impact levels, as delta (differential) obtained by comparing against a benchmark.
- *Risks associated with the Costs and Benefits*—the likelihood of a cost or a benefit item being realized, based on the confidence associated with their value estimates.

Because of the law of diminishing returns, regardless of likelihood, it is very difficult to eliminate risks. The benefits gained would not improve dramatically even with the expenditure of significant effort. Inversely, risks that can be reduced with minimal effort usually make the agenda of decision-making easy. Rather than assuming the expected safety factor assumptions would do, calling a design consultant to learn about the exact safety factors used in a structural design is worth the three minutes that it normally takes to consult the engineer.

For events in between, we rely on benchmarks. We propose that, if it has the following properties, a post facto case can be a good benchmark for a pre facto analysis situation:

- *Architect*—similar in stature and impact on the field, style not withstanding
- *Patron*—similar wealth, goals for project and determination to complete it
- *Environment*—of a similar scale, urban-ness, and environmental factors
- *Culture and media*—same national or cultural context, similar type of media
- *Commercial products*—similar commercialization trends and technologies
- *Political establishment*—similar political system and influence of political institutions

Proper judgment must be exercised to determine that the post facto data comes from a case similar to the current case in at least 70% of the aspects of each case. At a minimum this comparison must be leaning clearly towards similarity even though it may not be equivalent to identity. The critical step here is to be able to estimate the coefficients of risk for both cost and benefit, since these risk factors are likely to be significant due to the fact that many estimated values need to be carried over from benchmarks. So let us devote some attention to risk estimation here.

Post facto Analysis: Archival Documentation of Parameters

This approach is used when there is an abundance of data available on the critical aspects of a VBD project like the Swiss Re (Chap. 10). These aspects include, but are not limited to, project size, scope, client, site, architect, cost, schedule, and facility program. At a minimum, the project must be well documented in archival records, in terms of VBD features, including the costs and risks taken in realizing these features, and the benefits accrued because of these features. This approach is particularly effective for post facto analysis of the VBD since this is the only accurate way of measuring the features that must be included for a positive DAV assessment. To carry out the Archival Documentation of Parameters, the same procedure used in the case of pre facto analysis (above) can be used.

Documenting the Features

In Chap. 3 we provided four key features for Fallingwater: (1) cantilevers, (2) novel arrangement of physical elements such as window placement of mullions at corners of rooms, (3) vertical chimney and wall elements visually and structurally counterbalancing horizontal cantilevers, and (4) stone ledges forming a part of the hearth of the fireplace and cascading ledges of the terraces.

Let us consider the cantilever feature as an example. What is included in this feature and what is excluded? Is it just the floor area, or does it also include volume? Is it the structural component, or all of the architectural finishes that go with it? Is this mutually exclusive with the other features like the vertical elements, mullions, and so on? These questions would receive different answers depending on the frame within which they are viewed. Viewed through the lens of VBD, a feature is seen as the difference between having the feature versus not having it, i.e., a conventional solution, such as load-bearing construction.

As the benchmark, the cost of cantilevered construction can be compared against standard construction of the day. Table 7.4 indicates a format that can be used to encode the features and the benchmarks that are associated with them. This table is filled out for Fallingwater and as single rather than itemized features. Depending on the case, each feature can be broken down into finer grain sub-features, such as first, second, and third floor cantilevers, instead of "cantilever structure."

Documenting the Stakeholders

Features can result in costs and benefits that affect the various stakeholders differently. While the features described in Table 7.4 matter a great deal to Wright, the architect, and his cohorts and students, it does not matter to the client or the local

Table 7.4 Feature and benchmark descriptions in *fallingwater*

#	Feature	Description	Benchmark	Description
1	*Cantilever structure*	Three floors of 10–17 feet unsupported extension of floor-slab over the waterfall	*Load-bearing structure*	Perimeter walls and interior load-bearing elements are carried straight to foundation
2	*Corner mullions*	Novel elements: Place window mullions at corners of rooms	*Regular mullions*	Windows are placed in the body of walls not corners
3	*Vertical elements*	Chimney and walls counterbalancing horizontal cantilevers	*Walls and chimneys*	Walls and chimneys are placed in relation to the internal functions
4	*Natural elements*	Stone ledges forming part of the fireplace hearth and the terraces	*Placement on site*	The building is placed across from the falls on bear run creek

community what specific feature is used so long as it distinguishes the building as a novel entity. Consequently, it is of paramount importance to evaluate value of these features for each stakeholder.

Document Costs at Three Levels of Impact

Cost items are indicated as singular aggregate amounts for each feature (Table 7.5). This is suitable in cases when the feature cannot be broken down into sub-features. When this is possible, the cost of the feature can be shown as the sum of the costs of sub-features: $\Sigma^{k=n}$ $[a_1, a_2, a_3, \ldots a_k]$. This is to enable the expression of different granularities for the cost–benefit estimation. Table 7.5 also illustrates the case when the estimate for each cost item is a range of values rather than a single number. As an illustrative example, let us consider the *immediate impact* of *cost* for the *owner-client* stakeholder. Other categories are left up to the reader to complete a table for as indicated in Table 7.5.

If the building were to be built without its remarkable features, clearly it would cost less. In 1935, the original budget for the project set by E. J. Kaufmann was in the range of $20,000-30,000. This escalated to $70,000 in the first stage of the construction, which did not include the guest house, garage, pool, and servant's quarters. Eventually, when all was completed including the landscaping around the building, the cost was $250,000.

A rough estimate of the cost of the VBD features of the house—including the cantilevers as opposed to conventional construction of comparable square footage, the corner mullions, the stonework and its details, the interior finishes and details to complement the horizontal divisions of the massing and so on—would be the difference between the original cost minus the budgeted amount, or $40,000. This is a 133% increase over the original budget.

A more accurate cost for the VBD features could be estimated by breaking down each feature to sub-features and comparing them to corresponding, "normal,"

Table 7.5 Cost–benefit calculation of features

Feature	Estimate	Benchmark	Estimate
Cost with features	*A or* $\Sigma^{k=n}$ $[a_1, a_2, a_3, \ldots a_k]$	*Cost of benchmarks*	*B or* $\Sigma^{k=m}$ $[b_1, b_2, b_3, \ldots b_k]$
Cost of features	$B - A = C$ *or* $\Sigma^{k=n,\,k=m}$ $[b_1\_a_1, b_1\_a_2, b_1\_a_3, \ldots b_k\_a_k] = C$	*Caveat*	An option costs *D*; cost of feature should be in the range *C* to *(B – D)*
Benefit with features	*E or* $\Sigma^{k=n}$ $[e_1, e_2, e_3, \ldots e_k]$	*Benefit of benchmarks*	*F or* $\Sigma^{k=n}$ $[f_1, f_2, f_3, \ldots f_k]$
Benefit of features	$F - E = G$ *or* $\Sigma^{k=n,\,k=m}$ $[f_1\_e_1, f_1\_e_2, f_1\_e_3, \ldots f_k\_e_k] = G$	*Caveat*	An option yields benefit of H; benefit of feature should be in the range *G* to *(F – H)*
Cost–benefit summary	*Net benefit* $= G - C$	*Caveat*	Range of benefit = *(G – C)* to *((F – B) – (H – D))*

construction benchmarks. The difficulty here is to find truly corresponding benchmarks. What would be the corresponding construction element for the cantilevers? Would that be load-bearing structures placed on the falls or construction across from the falls, as Kaufmann first envisioned his cottage in the woods would be? In the current discussion we will be content with the approximate version of the estimate.

Document Benefit at Three Levels of Impact

Benefit items are indicated as singular aggregate amounts for each feature (Table 7.5). This is suitable in cases when the feature cannot be broken down into sub-features. When this is possible, the cost of the feature can be shown as the sum of the costs of sub-figures: $\Sigma^{k=n}$ $[e_1, e_2, e_3, \ldots e_k]$. This is to enable the expression of different granularities for the cost–benefit estimation. Table 7.5 also illustrates the case when the estimate for each benefit item is a range of values rather than a single number. As an illustrative example, consider the *immediate impact* of *benefit* for the *owner-client stakeholder*. Other categories are left up to the reader to complete a table for as indicated in Table 7.5.

E. J. Kaufmann anticipated that this venture would provide publicity for his commercial establishment that would eventually translate into increases in the annual balance sheets. He even put explicit effort into generating the "spin" around his house. Unpredictably, he had no control over the naming of the building, which made it impossible to reap the benefit of the "buzz" that he helped create. However, due to yet another set of unpredictable events, namely the passing of Mrs. Kaufmann, underuse of the property, and eventually the bequeathing to the Western Pennsylvania Conservancy by Edgar Kaufmann, made the association with the family name much more prominent.

Ironically, now that the property is firmly associated with the Kaufmann name, the commercial entity Kuaffman's Department stores, having been bought by Macy's, no longer exists. This leaves us with very little to show for in terms of measurable gains, other than the tax write-off realized by Edgar Kaufmann when he turned it over to the Western Pennsylvania Conservancy, and the opportunity gains of the funds saved from the annual maintenance costs of the building. These are a matter of semi-public records and therefore can be accurately estimated.

Exercise 6

Select a case study from among those presented in this text. Determine the primary functions and associated costs for it (Fig. 7.1). Determine if the costs are accurate. Next, find the distinguishing features of the building that correspond to primary functions. Note that some primary features can add new functions and new costs (Fig. 7.2). Discuss if your analysis of function-cost pairing is a more accurate reflection of the DAV of the building than the function-feature-cost set. Explain why.

Chapter 8
Quantitative DAV Analysis Methods

In this chapter, we will review a number of techniques for quantitatively supporting DAV analysis including scaling ordinal values, risk analysis, Bayesian probability, sensitivity analysis, and optimization.

Scaling Ordinal Values

The cantilevered form of Fallingwater is central for its innovative features. Earlier we estimated the cantilevered form of Fallingwater's construction costs. Yet, the benefits, tax deductions for Edgar Kaufmann or the NPV of the property are at best speculative matters. Measurement of the VBD parameters even when there is adequate data is a challenge. What would be the sales value of the property, today? Currently, who is benefiting from it? Did it enhance Edgar Kaufmann's career as an architectural historian at Princeton University? Did it make his and other's insights into architecture sharper? If so, how can we possibly measure benefits of this kind that are difficult to express in financial terms? Are these the only benefits we can cite?

In this case, comparative assessments placed on an ordinal scale can be very useful. Let us take the current value of the property. We know that in addition to the initial costs, the building has gone through $6 million preservation project starting in 2002. We can safely say that the asset value of the property is in excess of $5 million. Would the Western Pennsylvania Conservancy be willing to sell it for such an amount? This may depend on the revenues they are currently collecting. Given that Fallingwater has placed the Western Pennsylvania Conservancy on the map and is the most visited residential example of architecture in the USA, it would not be for sale any time soon even for such a large amount. However, for the purposes of DAV analysis we can establish an ordinal value, say $25 million, for the current worth of the property.

Ö. Akın, *Design Added Value*, https://doi.org/10.1007/978-3-030-28860-0_8

Likewise, on a dimension of career enhancing qualifications, its benefit to Edgar Kaufman can be placed between, say, the worth of earning a Master of Architecture degree and that of earning a PhD in Architectural History. This is an assessment that can be made accurately only by Edgar Kaufmann. However, this example serves to illustrate that benefits, when immeasurable in absolute terms, can still be included in DAV analysis as an ordinal value to quantify qualitative dimensions.

Simple strategies can be used to construct ordinal scales for assessing the worth or probability of events occurring in the construction of buildings. A five-point ordinal probability scale such as *certain, high, medium, low, unlikely*, while inadequate for fields requiring high accuracy, can be perfectly suitable for DAV analysis. In cases where these qualitative assumptions are included in quantitative calculations, straightforward mapping functions can be used; for instance: Certain ~0.99, High ~0.75, Medium ~0.50, Low ~0.25, and Unlikely ~0.01. Such mappings must always be accompanied by a sensitivity analysis, particularly if the evaluation of the alternatives is very close. In these cases, greater range of quantitative values should be assigned and values recalculated.

Sensitivity Analysis of Parameters

Sensitivity analysis is the study of how uncertainty in the output of a model (quantitative or otherwise) can be apportioned to different sources of uncertainty in the model input. A quantitative model is defined by a series of equations, input factors, parameters, and variables aimed at characterizing the process being investigated. Input is subject to many sources of uncertainty including errors of measurement, absence of information, and poor or partial understanding of the driving forces and mechanisms. This uncertainty imposes a limit on our confidence in the output of the model.

There are several possible procedures to perform sensitivity analysis. The most common sensitivity analysis method is sampling-based in which a calculation is executed repeatedly for combinations of values sampled from the distribution of the input factors. Sampling-based methods can also be used to decompose the variance of the model output. The sensitivity analysis process may have the following steps:

1. Specify the target function and select the input of interest.
2. Assign a distribution function to the selected factors.
3. Generate a matrix of inputs with that distribution(s) through an appropriate design.
4. Evaluate the model and compute the distribution of the target function.
5. Select a method for assessing the influence of each input factor on the target function.

In models involving many input variables, sensitivity analysis is an essential ingredient of model building and quality assurance. National and international agencies involved in impact assessment studies have included sections devoted to sensitivity analysis in their guidelines. Examples of such agencies include the European Commission, the White House Office for Budget and Management, the Intergovernmental Panel on Climate Change, and the US Environmental Protection Agency.

Risk Analysis

The concept of risk in decision-making is less of a notion of hazard and more of one of event probability. When applied in cost–benefit analysis, *risk* indicates likelihoods of the cost or the benefit event occurring at the expected level (or value). For example, are the sales receipts at the set value going to meet the minimum amount predicted? Or will the cost of energy used by the project be sustainable over the amortization period, or life cycle of facility? These are estimates of risk for alternative choices considered by the VBD decision maker.

Every time a VBD decision maker choses an available option, she runs the risk of making the wrong one. Converting this notion into a general theory of decision-making requires that the risk in making decisions be measured, regardless of how mundane or hazardous the consequences might be. Since every measurement runs the risk of being incorrect, the aspect of hazard or the lack of it then can be represented as coefficient representing the probability of uncertainty or inaccuracy of an estimate, to be multiplied by the value of cost or benefit. In other words, the estimate is calculated by taking the product of likelihood and expected value.

In design decision-making we need to consider various risks: likelihood of a client accepting a design, the likelihood of the engineer completing the structural design by a deadline, the likelihood of the structure collapsing before its fifth year of occupancy, or the likelihood of getting paid adequately for a design. To codify the risk of costs and benefits, quantitatively or qualitatively, two ingredients must be estimated: likelihood (probability) of an event, which is a VBD feature, and its importance (value). For instance, the next time we build a building like the Swiss Re, we may choose a different site that has fewer urban and historic challenges.

Utility theory provides an approach that can guide estimation of costs and benefits in decision-making. Economic utility is based on the assumption that personal or societal choices between alternatives can be measured by the values that constitute indifference of decision makers to the choices that they have. These values, plotted as indifference curves (dashed lines in Fig. 8.1), indicate positions where choosing for likelihood of occurrence and severity of occurrence of an event are shown in combination (blue dots, C1, C2, …, in Fig. 8.1). For instance, choosing the Diagrid structure over post-beam structure presents a likelihood of realization as a function of costs and benefits, as well as severity of the outcome also in terms of costs and benefits. Is it possible to realize the option and what would be at risk? Given these values, it is feasible to determine whether it is possible to realize this option and what the associated risk would be. The choices, C11, C6, and C7, all on the same iso-risk curve, present equally valued risks of realization. In other words,

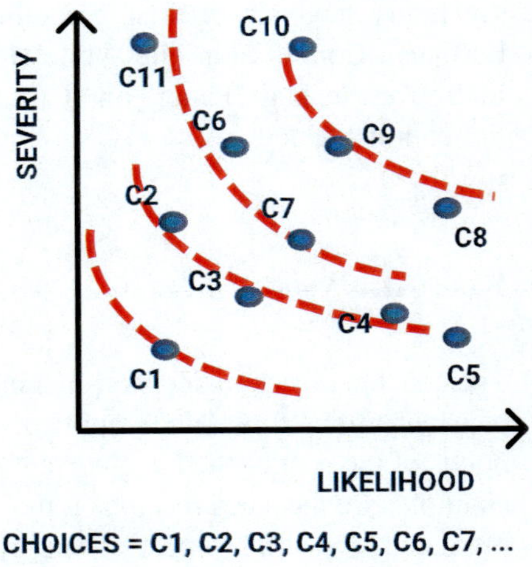

Fig. 8.1 Indifference curves for choosing between alternatives

choosing one of these against any other choice on the same iso-risk curve would be an equal bet. Whereas moving on to another choice on a different curve would present a different risk value. Iso-curves help organize choices along a given criterion, in this case "risk," to help decision-making. The critical aspect of each of these examples lies in the degree of confidence we can assign to each judgment.

For instance, an architect has as good or as poor a chance (therefore, 50–50) of convincing a random client about the goodness of her design. This is not because she knows the odds to be so but because she does not know anything about the clients' likes and dislikes at the onset. If she did know the client's likelihood of preferring a design over another, Bayes Theorem would apply. Hence, sometimes, risk assignment needs to be made even in the event that information is incomplete. In such cases we can make assumptions based on general trends.

Deadlines by design consultants are met most of the time (about 70%). This is a rough estimate of a marginal probability, an opinion that may or may not hold in a specific case. However, poll takers and statisticians would assure us that given a sufficient number of tries this would be a reliable planning benchmark. In the case of the Citicorp building, for example, we know that after its construction the likelihood of disaster hitting was estimated to be about 16% each year it was spared this disaster. This was a matter of eventual certainty. And finally, as subjective as it may sound, architects universally value their work more than what they earn in monetary terms.

It is critical to evaluate risk when exact numerical values are not at hand. In such cases we estimate using ordinal values or even qualitative descriptors based on past experience. Thus, another way to undertake cost-benefit analysis is to view it in terms of certainty. When the probability of an event K is so small that it is negligible or so high that it is unavoidable, we say that there is "certainty." When there is uncertainty, the value of risk must be somewhere in between. Uncertainty implies

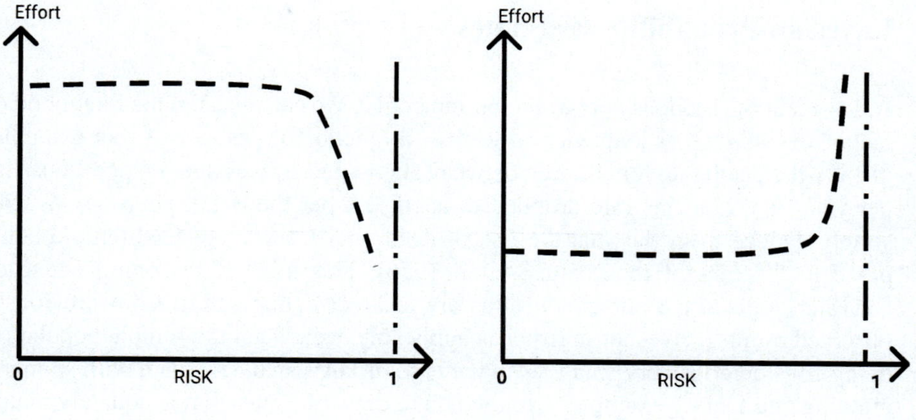

(a) (b)

Fig. 8.2 Risk prone behavior

risk, and when there is risk, it must be accommodated. This involves reducing or eliminating uncertainty, consequently risk. The context of the risk—perception of those subject to risk, criticality of the problem at hand, and availability of information about the case—is a key determinant of how uncertainty can be reduced.

Some individuals are risk averse, where even the smallest risk must be eliminated (Fig. 8.2a). Some are risk prone; they ignore uncertainty until it becomes an unavoidable (Fig. 8.2b). Critical problems require risk averse behavior. Sending humans to space, eliminating global change, or running nuclear plants are all critical tasks. Taking the bus, making some tea, or watching TV (normally) is not. In some cases, the perception of those involved in decision-making is biased to treat certain noncritical tasks as if they are critical. More often than not, this is so when human well-being especially of those individuals directly involved in the decisions is impacted: turbulence in the airplane, exposure to trace radiation, or estimating the strength of load-bearing structures.

A key decision in cost-benefit analysis for VBD is how much effort or resources should be dedicated to risk mitigation. The importance of solving a problem of risk reduction can be quantified as the magnitude of risk probability combined with the magnitude of benefit due to the reduced risk. When very large, this magnitude can become a deterrent for risk reduction. Likely events, like the intentional demolition or long-term deterioration of a building, are best left outside of the scheme of risks to avoid. Inversely very unlikely events, like a 500-year hurricane or fuel loaded jumbo jet aircraft colliding with a building, do not constitute realistic design criteria. Some methods in DAV analysis that help in assessing risk include Bayesian statistics, benchmarking, and sensitivity analysis.

Bayesian Probability Estimates

Assessment of risk does not have to be subjective. We can measure the likelihood of factual events such as longevity or death as a function of past events. For example, in 1900 the death rate for children between ages 5 and 14 was about 4 per thousand, per year. In 1975, this rate dropped to about 0.4 per thousand, per year. We can simply deduce from this that the risk of death in 1975 was reduced tenfold compared to the turn of the century (Doll 1976). This kind of reasoning underlies Bayesian logic since our conclusions are a direct extrapolation from previously observed events. Bayesian statistics is applicable even when the sample population being considered is very small, as in the case of DAV analysis. Even with so many building and facility examples from which to draw, the ones that adequately match a given VBD feature and, more importantly, have sufficient archival information to substantiate the VBD parameters is usually no more than a handful.

Bayesian probability defines probability as the degree to which a proposition is believed to be true. Starting with such a belief we can update this "degree of truth," or likelihood, in the light of new events that are relevant to that truth. Bayes' Theorem relates the conditional and marginal probabilities of current, say *X*, and future, say *Y*, stochastic events:

$$P(X|Y) = P(Y|X) * P(X) / P(Y) \tag{1}$$

where P(X) is the marginal probability of X;
P(X|Y) is the conditional probability of X, given Y;
P(Y|X) is the conditional probability of Y given X;
P(Y) is the marginal probability of Y.

To put it in lay terms, consider the following conditions: if the chances of winning a football game (marginal probability of *X*) is 50% for a team that usually wins half of their games; and the chances that they will beat the next team (marginal probability of *Y*) is 75% for an opponent that loses 25% of their games. The conditional probability of *X* given *Y*, when the first team's record against the second team this season is 60%, then is calculated as:

$$P(X|Y) = 0.6*0.75/0.5 = 0.9 \text{ or } 90\%.$$

Optimization with Graphics Methods

Design in general is not a problem of mathematical optimization. In fact, some design problems are often referred to as wicked problems as they hold all components of often changing contractual and requirement obligations. However, mathematical optimization perspective is useful to supplement design decision-making because it supports quantifying the trade-off analysis that is the underlying essence of design decision-making.

Optimization in engineering problem-solving is used to determine the best solution among all possible solutions in a given problem set under given set of top priority constraints. That is, techniques of optimization are directed towards the task of describing and selecting from among alternatives as opposed to the generative methods described earlier. In generative layout systems, the objective of optimization problems is to specify solutions in ways that satisfy predetermined constraints.

In optimization, problem formulation has three distinct ingredients: *variables, constraints,* and *objective function(s)*. Any design problem that lends itself to this kind of representation can be formulated as a problem of optimization. Under normal[1] conditions, once such a problem description is obtained, standard techniques of optimization can be used to find the solution that will assign values to the variables so that the constraints are not violated and the objective function is satisfied. Since the parameters of the optimization problem can be mathematically defined, its evaluation is quantifiable.[2]

Building design problems are inherently difficult to formalize[3] and quantify, especially in their entirety. While solving them, designers liberally define and redefine their design problems (Akin et al. 1992). This is a function of the partial-solutions designers obtain in order to integrate marginal sub-solutions into a whole. In such problem domains, called *ill-defined* by some, optimization formulations are difficult to construct. Radford and Gero (1986) pose the following feasible formulation:

> *Imagine an office block with continuous perimeter windows and a continuous horizontal sunshade above those windows... Imagine, too, that everything about the building is fixed except the height of the windows and projection of the sunshade. This gives us just two design variables to manipulate: we can combine different values of window height with different values of shade projection to yield tall, medium or short windows with extensive, medium or stubby shades... The windows cannot extend higher than the ceiling, and the sunshades are subject to a constructional constraint that restricts projection to a maximum of, say, 2 meters or so. These constraints together define a feasible region within which any design must lie.*[4]

In the authors' own words, this problem is "*simple and rather artificial.*" This is due to the fact that in this design problem there are many other relevant, but difficult to formalize variables, such as glazing type, depth of sill, functional requirements of a room, and standards of lighting levels. However, there are many circumstances, such as the one proposed by Radford and Gero, under which a complex design

[1] By "normal" we mean the presence of mathematical formulations of all of the three ingredients and absence of "incomputable" domains of optimality.

[2] It is conceivable that one can represent a problem as an optimization problem without using mathematical notation. In such cases the standard techniques of evaluation do not apply and there is little guarantee that this kind of formalization is of any use.

[3] In the mathematical sense.

[4] Radford A, Gero J (1986) Design by optimization in architecture and building. Van Nostrand Reinhold, New York.

problem can be structured as an optimization problem. The solution to this problem should consist of a set of *design values* to be assigned to a set of *design variables* subject to an *objective function*.

If we were to assign a value for the height of the window, say d_w, and a value for the depth of the overhang, d_o, this would be a step towards a solution (Fig. 8.3). For it to become an acceptable solution, these values, $V(d_w)$ and $V(d_o)$, would have to conform to the limitations set by the room height, facade orientation, and overhang cantilever dimensions. For this potential solution to be acceptable, the performance expectations implied by the solution, in terms of room illumination and summer solar heat gain, would have to be within desirable ranges.

Whether the client's objective is to minimize gain, to maximize illumination, or both, these requirements must be satisfied. Thus, we need to understand the relationship of the values that constitute the solution with respect to both the *design constraints* (Fig. 8.3b–d) and the performance objectives (Fig. 8.3e, f).

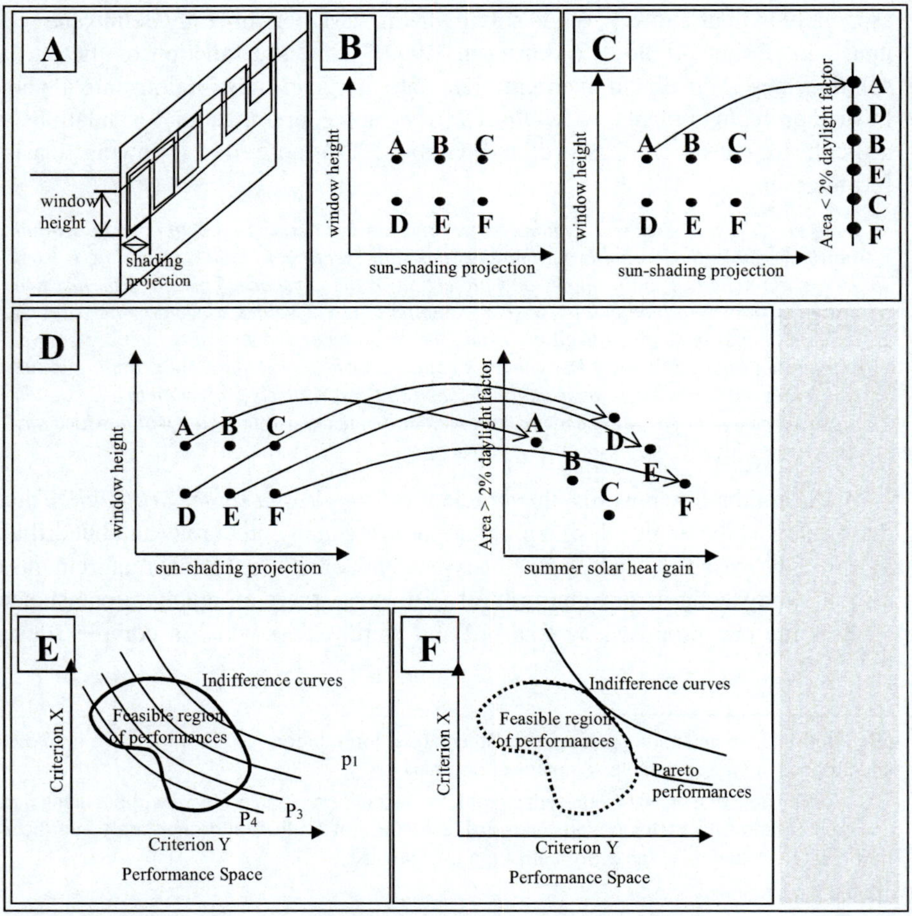

Fig. 8.3 Optimized window design (Radford and Gero 1986)

Optimization by Methods of Calculus

So far, our example solution in Fig. 8.3 is limited to a graphic approach. While visually obvious and intuitively clear, such approaches do not easily scale up to more complex optimization problems. One reason for this is the inherent difficulty of searching within the graphic space where a quantifiable, global metric, indicating the likelihood of success of a given move in the search space, does not exist. Another one is the difficulty of specifying an objective function, such as minimizing or maximizing one of the design values that cannot be graphically represented. A third difficulty is the multidimensionality of optimization problems that cannot be easily mapped into the two-dimensional space of graphs. Of course, there is also no guarantee that the graphic representation will include the globally optimal solution.[5] Methods more robust and accurate than the graphic ones are also available, such as differential calculus, in which the derivation of the optimal solution from the initial problem definition is carried out strictly mathematically.

Let us select a problem to make a strictly mathematical solution description more concrete. Assume that in designing the Kimball Art Gallery, Louis I. Kahn has just decided to make a vaulted gallery module. As history would attest, this would be a semicircular vault supported by end walls and side beams (Fig. 8.4). This eliminates the sidewalls, as the vaulted module attaches to the next one along the bean on its broad side. The section of each module below the vaulted space turns out to be a perfect square. Let us add a few other constraints to make the problem more interesting. Assume that the mechanical system selected for the project dictates that the volume of each module should not exceed 20,000 cf. In the event that this must be exceeded for reasons outside of mechanical considerations, the next tier of constraints turns out to be 30,000 cf. The optimization task then is to determine the dimensions of the module so that its overall cost is minimized.

The initial representation of this problem is shown in Fig. 8.4a and its mathematical formulation that is represented by the following steps is shown in Fig. 8.4b:

1. Find an expression for the areas of surfaces to be built.
2. Find an expression for the module's volume, based on 1.
3. Find an expression for the module's overall cost, based on 1.

The solution to the problem is shown in Fig. 8.4c and contains these steps:

1. Set volume equal to 20,000 cf. as required.
2. Find an expression equal to one of the variables, such as length of the module (L), in terms of the other, that is the radius of the vault (R).
3. Substitute one expression into the other, reducing the number of unknowns in the expression to one (R).

[5] Basically this is the distinction between *optimization* and *pareto optimization*, which is also known as sub-optimization, local optimization or satisficing.

A graphic representation of the optimization problem

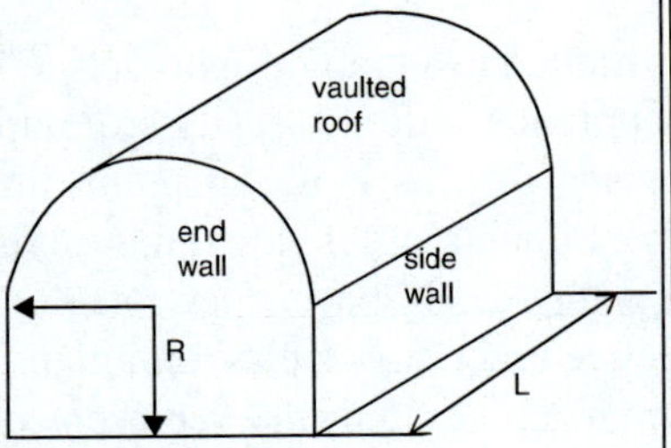

Roof Cost = $15/sf
Endwall Cost = $30/sf
Sidewall Cost = $0/sf
Volume = 20,000 cf
Minimize Cost (objective function)

B Formulation of the optimization problem

Roof area $= \pi RL$

Endwall area $= 0.5\pi R^2 + 2R^2$
$= (2 + 0.5\pi)R^2$

Volume $= (2 + 0.5\pi)R^2L$

Cost $= 15(\pi RL) + 60\,(2 + 0.5\pi)R^2$
$= 15\pi RL + (120 + 30\pi)R^2$

C Solution of the optimization problem

$(2+0.5\pi)R^2L = 20{,}000$
$L = 5{,}600/R^2$
$C = 47.12\,R\,(5{,}600/R^2) + 214.24R^2$
$= 214.24\,R^2 + 263.872/R$
$dC/dR = 428.48R + (-263.872)R^{-2} = 0$
$428.48R^3 = 263.872$
R = 8'6";
L = 77'6";
C = $46,522.60
Gallery width = 17'

D Design implications

The width of the space is too narrow (17") for a gallery space, therefore, let volume = 30,000 cf, and height = 2/3 R

E Formulation of the revised problem

Roof area $= \pi RL$

Endwall area $= 0.5\pi R^2 + 1.33R^2$
$= (1.33 + 0.5\pi)R^2$

Volume $= (2 + 0.5\pi)R^2L$

Cost $= 15(\pi RL) + 60\,(1.33 + 0.5\pi)R^2$
$= 15\pi RL + (79.8 + 30\pi)R^2$

F Solution of the revised problem

$(1.33+0.5\pi)R^2L = 30{,}000$
$L = 10{,}330/R^2$
$C = 47.12R\,(10{,}330/R^2) + 214.24R^2$
$= 214.24\,R^2 + 486{,}760/R$
$dC/dR = 428.48R + (-486{,}760)R^{-2} = 0$
$428.48R^3 = 486{,}760$
$R^3 = 1{,}136$
R = 10'5";
L = 94'11";
C = $70,029
Gallery width = 21'

Fig. 8.4 Optimized solution for the vaulted space design problem

This yields an expression that states the cost of the project in terms of the radius of the vault (Fig. 8.4c). Since this is a quadratic function, differentiating this expression with respect to the radius and setting it equal to zero can yield the minimum or maximum points on this function.[6] Thus, we find that R = 8′ 6″, and L = 77′

[6]We know from calculus that any tangent to a function is defined by the derivative of the function. By setting the derivative equal to zero and solving for the value of the variable in the expression we find those points where the slope of the tangent is equal to zero, i.e., minimum and maximum points. To find whether this point is a minimum or a maximum we can check the sign of the second derivative. If it is negative, then the point is a maximum since the rate of change in the slope of the tangents at a maximum is diminishing; and minimum if it is positive.

6″. This ensures that the volume of the resulting design will not exceed 20,000 cf.; the section of the module will fit perfectly into a square and the estimated cost of each module will be $46,522, which is the least cost under the constraints specified.

This is a truly unique optimal solution that satisfies all of the stated constraints while minimizing costs. Any other set of values for R and L will result in costs that excess this one. Yet, it is quite conceivable that this may not be the preferred solution for a variety of reasons. It is conceivable, for example, that the overall width of the module (17′) would not be suitable for the exhibition function due to a number of problems: circulation, lighting, and size of exhibits (Fig. 8.4d). At this point, the designers may want to alter some of the initial assumptions of the formulation.

To illustrate the point, Fig. 8.4d shows the way the problem can be reformulated to remedy this concern: (1) update the height to radius ratio to 2:3, and (2) move up to the next volumetric threshold of 30,000 cf. Figure 8.4e and 8.4f show the updated problem formulation and its solution, respectively, following the same steps followed earlier. This yields an increase in the width the module to 21′ achieving the desired effect of the reformulation.

Optimization by Simplex Methods

Not all problems of optimization can be formulated by graphic or methods of general calculus. Some problems involve so many variables that either method may become difficult or impossible to use. In these cases, a computational technique called the Simplex Method can be used, where

1. The variables are continuous and take values equal to or greater than zero.
2. The constraints are a linear function of the variables.
3. The objective function is a linear function of the variables.

The operations of the simplex method are in essence analogous to the graphical method. However, they are mapped into the numerical domain so that optimal solutions can be estimated using algorithmic devices, such as matrix algebra. We recommend that the reader to consult one of many texts in this area for further exploration (Radford and Gero 1986).

Using optimization is the closest that design decision-making ever comes to a rational modality. And as our examples demonstrate, the application areas of optimization techniques are those where engineering aspects of design decisions dominate, such as that of the optimization of a vault design. In order for an optimization technique to be applicable to a problem, however, it must be possible not only to express the major design alternatives quantitatively, but also to estimate the tradeoffs between them. The burden of success of this approach then is more in formulating a problem than in solving it. Computationally intractable problems notwithstanding, a successful formulation all but guarantees the finding of the best solution that there is, provided that there is at least one available.[7]

Exercise 7.1

Find realistic situations in the Capital Delivery Process where the prototypical behavior resembles *risk averse* behavior and where it resembles *risk prone* behavior. Explain in detail the reasons why that is the case.

Are any of the examples you considered in answering the risk averse and risk prone behavior section of this question fall into Indifference Curves condition shown in the next graph below? Explain if this is an ordinary or an unusual circumstance.

Risk Averse Behavior versus Risk Prone Behavior

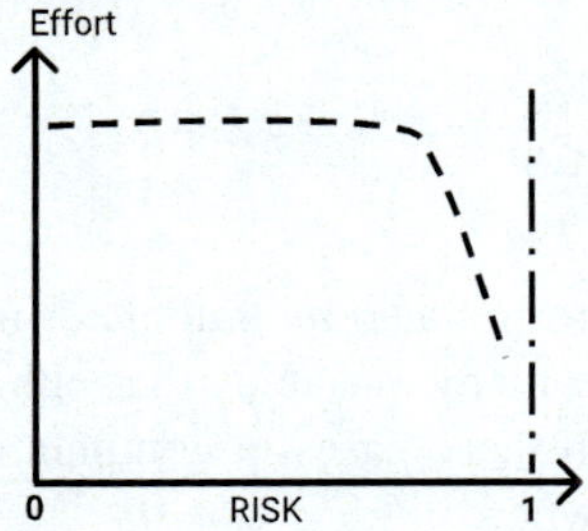

Risk Averse

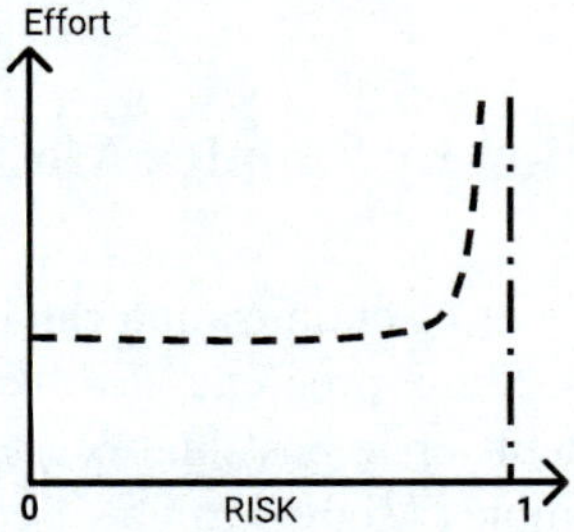

Risk Prone

Risk Analysis *Indifference curves* for choosing between alternatives

Indifference curves show equi-risk events, like the ones that fall on one curve [c1; c2, c3, c4, c5; c6, c7 c11; c8, c9, c10].

Equi-risk means the product of likelihood and severity is as close to others in the same risk curve are the same (or with in a predefined range).

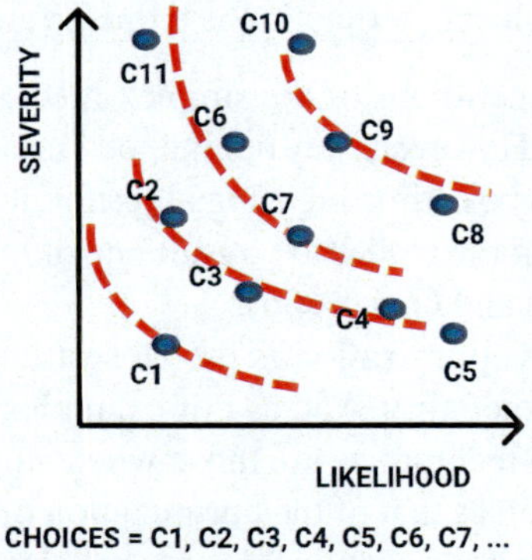

[7] This is not to suggest that there are no difficulties in the solution of optimization problems. In fact, there are many search spaces that are so large that solutions are not only difficult but impossible to find. We are not referring to these practical difficulties in the above statement.

Exercise 7.2

Benchmarking provides anchors or points of comparison for cost and benefit assessments for DAV features to make sure that these estimates are realistic. Use the table below to evaluate costs, benefits, and "caveats" applicable to design features of a building case found in this text.

Cost–benefit calculation of features with benchmarking

Feature	Estimate	Benchmark	Estimate
Cost of features	A or $\Sigma^{k=n}$ $[a_1, a_2, a_3, \ldots a_k]$	*Cost of benchmarks*	B or $\Sigma^{k=m}$ $[b_1, b_2, b_3, \ldots b_k]$
Cost of features	$B - A = C$ or $\Sigma^{k=n,\, k=m}$ $[b_1\_a_1, b_1\_a_2, b_1\_a_3, \ldots b_k\_a_k] = C$	*Caveat*	An option costs D; cost of feature should be in the range C to $(B - D)$
Benefit of features	E or $\Sigma^{k=n}$ $[e_1, e_2, e_3, \ldots e_k]$	*Benefit of benchmarks*	F or $\Sigma^{k=n}$ $[f_1, f_2, f_3, \ldots f_k]$
Benefit of features	$F - E = G$ or $\Sigma^{k=n,\, k=m}$ $[f_1\_e_1, f_1\_e_2, f_1\_e_3, \ldots f_k\_e_k] = G$	*Caveat*	An option yields benefit of H; benefit of feature should be in the range G to $(F - H)$
Cost–benefit summary	*Net benefit* $= G - C$	*Caveat*	Range of benefit = $(G - C)$ to $((F - B) - (H - D))$

Exercise 7.3

A farmer has 1500 ft of deer-proof fencing that she wants to use to create a vegetable garden at the back of her house. To fit in the backyard space, this area has to be a rectangular enclosure that provides the maximal area while having a width no less than half the length of the rectangle. What dimensions will optimally satisfy her requirements?

Chapter 9
Expertise, Innovation, and Creativity in Support of DAV

The quantitative analysis techniques we advocate to support DAV analysis are most successfully applied in the context of a design process that is iterative, responsive to change, and collaborative. Such a design process empowers creativity and innovation while anchoring design and its construction in fundamental quantitative engineering and economic trade-offs. Expertise, innovation, and creativity in design are essential in identifying features which drive DAV.

Innovation by eXtreme Design

It is likely that you have seen at least a little bit of eXtreme sports on television: young kids on roller-boards or dirt bikes, making loop-de-loops in midair without the benefit of safety nets. *eXtreme Design* takes its inspiration from the genre of things that are of that ilk. However, the intent of eXtreme Design is far from those of its sports cousins. It is not for the thrill or even the challenge that is involved. It is to respond to needs that are immediate which demand the kind of *agile* thinking, discipline, and acting that is required in most popular eXtreme domains.

eXtreme or Agile Design is not an entirely novel idea. Software engineers have been at it for a while (Ambler 2001). eXtreme Programming (XP) as a software development process advocates frequent releases of software, simple design, and timely reaction to changing requirements among other practices (Beck 2004). The agile approach is a philosophy for software design that emphasizes a new set of *values*, *principles*, and *practices*, rather than the prescriptive processes of the old. An abridged version of Agile principles and practices are provided below:

> *…Through this work we have come to* **value***: Individuals and interactions over processes and tools Working software over comprehensive documentation Customer collaboration over contract negotiation Responding to change over following a plan … We follow these* **principles***: Welcome changing requirements, even late in development. Agile processes harness change for the customer's competitive advantage. Deliver working software*

Ö. Akın, *Design Added Value*, https://doi.org/10.1007/978-3-030-28860-0_9

> *frequently, from a couple of weeks to a couple of months, with a preference to the shorter timescale. … Build projects around motivated individuals. Give them the environment and support they need, and trust them to get the job done. The most efficient and effective method of conveying information to and within a development team is face-to-face conversation…Simplicity--the art of maximizing the amount of work not done--is essential. The best architectures, requirements, and designs emerge from self-organizing teams. At regular intervals, the team reflects on how to become more effective, then tunes and adjusts its behavior accordingly.*[1]

The goal of agile development process in software engineering is to provide rapid and meaningful turnaround of design products by incorporating continuous feedback in the design process from the users and other experts. At a minimum, it provides an alternative to conventional and more cumbersome processes. The "Waterfall Model" of design, during which every stage of work must be completed before the next stage can be started, dominates the conventional approaches to software design. Such approaches have the distinct drawback of costly backtracking due to volatile requirements, particularly if the upfront design process takes a long time to complete.

Agile approaches do not run the same risks since they work with very rapid turnaround cycles. By the same token, products developed with agile approaches strive to be just "good enough," and creating a balance between anticipating all future requirements versus responding to what is high priority today. However, agile approaches also attract criticism as when a conscious focus on design is lacking changing requirements can also result in significant amount of costly rework. The quick delivery of interim products does not diminish quality standards for the final product. If anything, it means just the opposite. Rapid turnaround of design cycles means increased number of meaningful inputs from the client and design consultants, so that the quality of the final design is improved when agile is followed with the intended discipline it is created for.

Proponents of *eXtreme* programming advocate simple design as an extreme practice and argue:

> *We will continually refine the design of a system, starting from a very simple beginning. We will remove any inflexibility that doesn't prove useful. … Sometimes it is significantly more productive for a developer to draw some bubbles and lines to think through an idea, or to compare several different approaches to solving a problem, than it is simply start hacking out.*[2]

While the term may be new, the spirit of *eXtreme* Design is not new for building design. When the professional is asked to produce a complex building design within a matter of months, even weeks or days, she intuitively uses a form of *eXtreme* Design. Armed with this optimistic view, the *eXtreme* Design Manifesto in the domain of building design means: rapid change, quick delivery, individual motiva-

[1] The website http://www.agilemanifesto.com/ declares some of these values and practices.

[2] K. Beck and B. Boehm, "Agility through Discipline: A Debate," IEEE Computer, 36, 6, (June 2003): 44–46.

tion, face-to-face work, simplicity of expression, self-organizing teams, and self-evaluation.

In order to survive in the ever-changing world of design information, architects and engineers have to be self-organizing and self-critical. As the design solution takes shape, clients and contractors express greater tangible requirements to be met by the design. This, in turn, alters the design, causing an iterative refinement process to dominate the design activities. In fact, it would not be difficult at all to argue that architects and engineers are already trained to behave like Agile or *eXtreme* Designers that other professionals working in ill-defined problem domains[3] aspire to become.

Architects and engineers are accustomed to adjusting their designs to changes that emerge from evolving client needs and problem contexts, sometimes all the way through and even beyond the construction process (Akin and Anadol 1994). They can deliver design products as rapidly as the market requires them. Personalities involved in the design team are, more often than not, part of the intent behind team composition. Clients, consultants, and contractors do their best work, through negotiations about design decisions face-to-face. Simplicity of means to obtain the greatest effect with the least amount of resources is often expressed through the universal mantra: "Less Is More."

Creativity

A critical aspect of DAV analysis is to determine the design features that add value without inflating costs. In this analysis, we are particularly interested in identifying features that can take the value of the entire design to a higher level. This is a phenomenon that is commonly known as creative breakthrough or a Sudden Mental Insight (Akın and Akın 1998). Hence, we shall review the importance of decisions that often take place during the initial stages of design and manifest their impact on creativity through the sudden mental insight process. We will draw initially on building cases; our examples are from the creations of Antonio Gaudi. Next, we will propose a theory of creative insights based on findings about the Sudden Mental Insight process observed in inventions and puzzles.

[3] While the eXtreme design studio was my idea, I found many of my ideas resonate with those of Professor Hanson. In this book, I feel justified to take credit for "inventing" the eXtreme design approach; yet in its first incarnation, during May 16th through July 13th, 2003, in Praha, Czech Republic, both I and Professor Hanson took part in its implementation.

Early Stages of Design: Antonio Gaudi's Work

The practice of Antonio Gaudi, one of the most enigmatic figures from the turn of the nineteenth century, coincides with a confluence of transitions in the field of architecture and structural engineering. A significant feature shared by almost all of Gaudi's works is their spectacular expressions of form and unusual use of materials (Figs. 9.1 and 9.2). These superior works of architecture and engineering incorporate stylized organic forms and other enchanting images both in structural and decorative terms (Figs. 9.3 and 9.4).

It is difficult to say with any certainty how Gaudi came up with these designs. Such a task is difficult enough even with designers who design in the known and well-established styles and when their designs can be scrutinized directly. In the case of a designer whose profile fits more closely that of a genius engineer than an ordinary man, and some of whose works are being completed about a century after his passing, it is very difficult to say anything definitive about his design process. In fact, we would not even consider this case if it were not for some extremely interesting and valuable documentation which exists about Gaudi's "methods" of invention.

It has been widely documented that Gaudi's work aspires to the Gothic style in its unique way. Furthermore, his designs emulate the forms, colors, and textures of the physical environment of his native Catalonia. It is also widely known and documented that Gaudi's work style included the building of elaborate "tensile" structure models (funicular models) of his designs (Fig. 9.3). And there is strong evidence of

Fig. 9.1 Guel park (Image credit: "PA080675" by NUCO is licensed under CC BY 2.0)

Fig. 9.2 Casa Mila by Antonio Gaudi (Image credit: "Casa Mila" by iShot71 is licensed under CC BY-ND 2.0)

Fig. 9.3 Gaudi's funicular model

Fig. 9.4 Gaudi's cathedral interior design (Image credit: "No pensava veure mai aquest sostre finalitzat…" by SBA73 is licensed under CC BY-SA 2.0)

the direct use of the forms in many of his designs (Fig. 9.4). From these sources of evidence, it is possible to infer a pattern of reasoning, in however speculative terms it may be, that can shed some light upon his creative design process.

Gaudi employed his funicular models to create the vertically elongated forms of his designs that characterize many Gothic buildings. At the same time, Gaudi was interested in being unconstrained by orthogonal geometry that governed most formal aspects of masonry buildings. He was seeking forms that would only be governed by gravity. It is conceivable that he was also emulating the natural forms of the Catalonian landscape. His tensile experiments (Fig. 9.3) and the compressive structures derived from them (Fig. 9.4) followed the trajectories that gravitational forces constituted in these structures more closely than anything else.

It is sound inference from these observations that Gaudi restated the problem of design in a form that was different than those of his peers and predecessors. This was enabled by looking at the behavior of structural forces and by eliminating the "arbitrary" impositions of orthographic geometry; thereby, inventing a unique *Problem Restructuring* process.

Frames of Reference in Puzzles: A View of Expertise in Creativity

Our Sudden Mental Insight approach that has led to the insights about design creativity has been a phenomenon of managing the *Frames of Reference* that assist in solving routine problems more efficiently but become obstacles to creative solutions. Oddly enough some of the clearest evidence for this kind of human creative behavior is found in the area of problem-solving and puzzles.

The nine-dot puzzle is a task, which involves manipulations in the visual field. The puzzle is quite straightforward. The initial state involves an arrangement of nine-dots (Fig. 9.5a). The goal is to draw four straight lines so that each dot has a line going through it. Furthermore, the lines must be connected end-point to end-point so that only two end-points are not connected to another line.

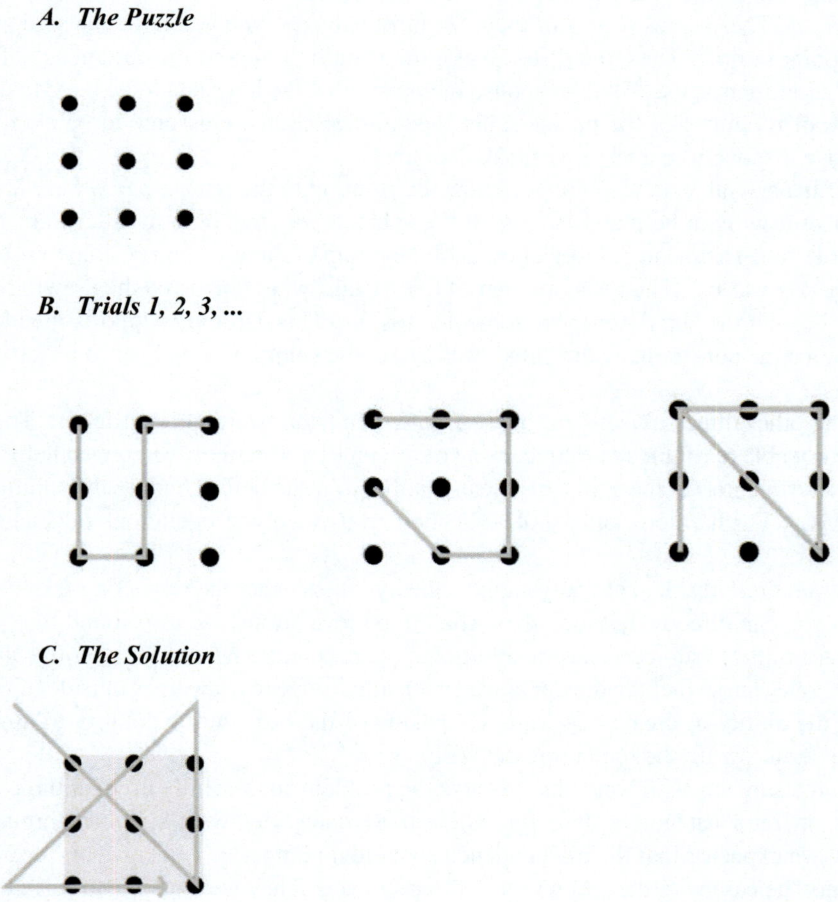

Fig. 9.5 The nine-dot puzzle, (**a**) initial state; (**b**) search states; (**c**) solution state

In order to successfully solve the problem, subjects must realize that they are allowed to extend a line beyond the square area formed by the nine-dots. Restricting the area in which lines are drawn only within the square space implied by the nine-dots renders the solution impossible. It is hypothesized that once this insight is attained, the solution to the problem is trivial. We conducted an experiment to explore the cognitive processes that lead to the use and reformulation of the Frame of Reference in solving this puzzle. Our results certainly supported the idea that this insight is essential in creating a problem space where the task can be solved.

First, subjects were read a standardized set of instructions explaining the task. Subjects proceeded until either they solved the problem or time constraints dictated a shift in strategy. In the latter case, subjects were given a series of hints or clues to improve their approach to the puzzle, in two tiers. In the first tier, they were given a partial solution diagram to help them find the complete solution. In the second tier, they were directed to add new dots beyond the square grid of the original puzzle diagram suggesting the parts of the solution that exceed the space defined by the nine-dots. These clues were sufficient for most subjects who had not done so up to that point in time. Thus, the discovery of the solution required the emergence of a revelation such as the "Aha!" response indicating that the lines had to go beyond the frame of reference of the puzzle. This sudden revelation we assume to be akin to creative discovery in a variety of task domains.

Of the six subjects who did not know the solution to the puzzle, three were able to solve it without help and the rest of the subjects required help in order to reach the solution. The mean number of trials of the group of three who were able to solve the puzzle was 6.7. The mean number of these trials, which involved the drawing of lines beyond the "box" for these subjects, was 5.4. This certainly supports the idea that once the constraint of the "box" was lifted, the solution turned out to be a trivial task.

The other three subjects' problem-solving behaviors were quite different. They were not able to lift the restrictive frame of reference before they were given help by the experimenter. In the end, their mean number of trials before finding the solution was 54.3. Furthermore, only 0.08% of their trials involved operations outside of the "box."

While these results generally support the hypothesis that the restricting frame of reference can effectively block the "Aha!" type revelations, we also found that, in the case of these subjects, several additional operations are needed in order to solve the puzzle. These include the operator, which allows one to draw lines outside of the box; the ability to create new vertices outside of the box; and the ability to align these vertices with the remaining dots (Fig. 9.5).

Three subjects who were able to solve the problem successfully used all three of these operator categories. In cases where these categories were supplied through hints, we expected that their dependencies would no longer be evident. This clearly was not the case with the second set of three subjects. They were unable to solve the puzzle when supplied a partial set of hints, suggesting the existence of dependencies between all three hint categories: the need to go outside of the restricting frame of reference, to create new vertices beyond the restricting frame of reference, and to align these vertices with those that are within the restricting frame of reference.

In conclusion, the hypothesis that going outside of the nine-dot space as the *single* additional new move-operator sufficient to solve this problem is not supported. It certainly is a necessary element; however, it must be coupled with the two additional move operators, the ability to create new vertices outside of the nine-dot space and the ability to align these vertices with the remaining dots.

Breaking the Frame of Reference

The commonly recognized "Aha!" response is a reference to the moment when a creative flash arrives. Famous examples of this creative flash include the discovery of penicillin by Sir Alexander Fleming and the Greek mathematician and inventor Archimedes' proclamation of "Eureka!" after formulating the theory of mass displacing its volume in liquids. While these are, at best, unsubstantiated anecdotes, they illustrate the phenomenon of the sudden onslaught of a novel idea, which is frequently experienced even under mundane circumstances: "why didn't I think of that before?"

If we examine other anecdotal, yet better documented accounts of creativity, we find a range of interesting issues to consider: Mozart speaks of the special qualities of the condition under which the creative impulse arrives.

> *When I am, as it were, completely myself, entirely alone, and of good cheer - say traveling in a carriage, or walking after a good meal, or during the night when I cannot sleep; it is on such occasions that my ideas flow best and most abundantly. Whence and how they come, I know not; nor can I force them. Those pleasures that please me I retain in memory, and am accustomed, as I have been told, to hum them to myself. If I continue in this way it occurs to me how I may turn this or that morsel to account, so as to make a good dish of it, that is to say, agreeably to the rules of counterpoint, to the peculiarities of various instruments, etc.*

Tchaikovsky notes that the arrival of the creative impulse is "sudden" and "unexpected." Generally speaking, the germ of a future composition comes suddenly and unexpectedly.

> *If the soil is ready – that is to say, if the disposition for work is there – it takes root with extraordinary force and rapidity, and shoots up through the earth, puts forth branches, leaves and, finally, blossoms.*

Both of these accounts confirm, at one level, the a priori impressions we have of the creative process. Yet, there are also new insights suggested. Tchaikovsky speaks of the "soil" being ready or the "disposition of the work" being there. Mozart speaks of retaining the memories of the pleasures that come and of mentally nurturing them into a "good dish." It is clear that the idea that arrives and what happens to it next are related. There is a readiness for the idea that is suddenly conceived.

While we will not be able to fully explain a process as complex as creativity by relying on just these metaphors, it is reasonable to infer that what arises so suddenly does not arise from nothing but from the cognitive preparation that anticipates and evokes the idea in the first place. Mozart in fact speaks about turning the "*morsel to account... agreeably to the rules of counterpoint, ...instruments, etc...*" There exists a body of rules by which the idea must abide.

This is neither a surprise nor an obstacle to the composer. If these were not anticipated at the conception of the initial idea, would they have been treated in such a matter of fact way? We argue not. In fact, we argue that the creative process is a whole in which the conception of the idea influences and is influenced by the anticipated development. Cognitive tools used in the development of the creative entity are also responsible for the "inspiration" that initiates the creative process.

In formulating our investigation of creativity, then, we consider that the sudden onset of a creative insight, which has eluded the composer or designer until that moment, is a key step. There is no doubt that the soil upon which the sudden mental insight germinates has to be properly and painstakingly prepared. In the case of the expert this is a laborious yet familiar process. But once it is complete, the conditions for the onset of the sudden mental insight are ready.

In the case of the nine-dot puzzle a similar alteration is required, which involves a direct manipulation in the problem representation. The net result in both cases is that a new construct consisting of both a specific problem representation and operations applicable to it must be created.

The fact that the creative process requires the discovery of a new problem space necessitates that it must be conducted in at least two distinct search spaces: one that of solutions and the other of ground setting problems. This means that the creative individual must have skills not just for problem-solving, but also for defining new problem spaces. This latter skill has been described in many different contexts and in different ways. One of the most important aspects of the process of searching for new problem spaces has to do with domain knowledge. As observed in puzzles, inventions, and designs, the knowledge of the creative agent plays a key role in their creative achievements.

Another relationship between these two forms of search is that discovering a problem space is based on the successes and failures of the process of searching for solutions in existing problem spaces, by measuring how close the earlier trials come to obtaining a solution. For example, the problem solver can adjust the parameters of a problem space to be closer to a solution domain. Consequently, these two search spaces are dependent on one another.

Finally, it is important to underscore the important role of the "Aha!" response as a related, but insufficient manifestation of creative acts. This seems to be more important for the sociopsychological aspects of discoveries and creative inventions, than for the cognition of creativity.

Exercise 8

You are given a mutilated checkerboard puzzle (two squares from opposite corners of the board are missing) and a supply of domino pieces. Each domino piece is just big enough to cover two adjacent squares on the mutilated-checkerboard. How many domino pieces will cover all of the present squares on the board exactly (that is, no squares can be left uncovered and no domino pieces can extend beyond the board)?

When you solve this puzzle describe the thought process you went through: Did you change your strategy of exploration? How did it improve or degrade your search process? Was there a moment of epiphany when you suddenly saw how this puzzle can be solved? Was this approach superior to the previously tried ones?

Part III
DAV Analysis of Two Seminal Case Studies

Previous chapters covered methods that constitute the analytical foundations of DAV, including cost–benefit analysis, data elicitation, pre facto and post facto analysis, scaling ordinal values, risk analysis, Bayesian probability, sensitivity analysis, benchmarking, optimization, and expertise-innovation-creativity in design.

In this Part, we include two detailed DAV analyses: the Swiss Re Tower in London (Chap. 10) and the Commerzbank Tower in Frankfurt (Chap. 11). Synopses of the methods used in their analysis are abstracted through a collection of information templates included in the Appendix section of the book, titled "*Design-Added Value Methods Recommended for Analyzing Case Studies*

Readers are encouraged to conduct their own detailed DAV analyses of new case studies in Chaps. 12 through 17 that present preliminary data about John Hancock Tower in Boston; K. C. Hyatt Regency in Kansas City; Pruitt-Igoe in St. Louis, MO; Crystal Palace in London; Sydney Opera House in Sydney; and Citicorp Tower in New York, using the templates provided in the Appendix.

Chapter 10
The Swiss Re Tower: Analysis of a Seminal Case

The Swiss Re Tower, designed as the headquarters of the Swiss Reinsurance Company, is situated at the historic core and financial district of the City of London. Designed by the architect Foster and Partners, it is one of the most distinguished and controversial office skyscrapers in London. Officially named "*30 St Mary Axe*," and popularly called the *gherkin*, the Swiss Re Tower adds a distinctive identity to the skyline of the city due to its form (Fig. 10.1). In addition, this award-winning building reflects Foster's ideals about quality of work space and Swiss Reinsurance Company's core values of sustainability (Powell 2006).

The history of the Swiss Re Tower project begins early in the 90s when the company decided to unite their five London offices in one building at the heart of the insurance district of London in order to incentivize interaction, interexchange of ideas, and creativity in corporate operations. Nevertheless, the process was not straightforward and the project became a reality in 2003, 13 years later. The main facts of final project are summarized in Table 10.1.

Project Parameters

The first constraint of the project, and perhaps its main challenge, was the site location. The company wanted to use the site of its building on the St Mary Axe Street, part of a Victorian redevelopment area occupied by historic medieval and first half nineteenth-century buildings. In April 1992, the city of London suffered an IRA (Irish Republican Army) attack at 30 St Mary Axe, killing three people and damaging several buildings, among which was the historic Baltic Exchange Company Building. Since 1987, the Baltic Exchange belonged to the Grade II list of the *Statutory List of Buildings of Special Architectural or Historic Interest*. After the attack, the City Corporation of London and the English Heritage organization ruled

Ö. Akın, *Design Added Value*, https://doi.org/10.1007/978-3-030-28860-0_10

Fig. 10.1 The Swiss Re Tower—30 St Mary Axe (Image credit: "cigar panorama Swiss Re Tower" by Rechanfle is licensed under CC BY-SA 2.0)

Table 10.1 Main facts of Swiss Re Tower (Munro 2010; Powell 2006; Morrin 2008)

Category	Facts
Type	Office Skyscraper
Location	The City of London
Design	Nov 1997–July 1999 (21 months)
Construction	2001–2003 (33 months)
Occupancy	Late 2003
Opened	May 2004
Site area	5667 m^2 (1.4 acres)
Gross floor area	74,300 m^2
Office space floor area	46,450 m^2
Retail space floor area	1400 m^2
Costs	\$215 million

that any future development on the site should restore the hall and rebuild the Baltic Exchange facade (Morrin 2008).

When the burden of the historical designation reduced the market value of the property, Trafalgar House bought the site for a mere £10.5 million and started the development of a building for Swiss Re. One year after, the City Corporation allowed Trafalgar House to explore the option of a "ground scraper" scheme that incorporated the Baltic Exchange facade.

In 1996, the English Heritage Head, Paul Drury, reported that the Baltic Exchange building was in worse condition than expected and a restoration would be a replication rather than a reconstruction. The demolition of the Baltic Exchange building was allowed in return for a high-quality development by a high-status architect. Two other factors supported this decision: (1) to keep the "square mile" area as a center of the financial and services activities of London, and (2) to develop areas outside of the corridors that protect the view of important historic buildings.

The second constraint of the project was to meet the City Corporation and English Heritage's height and shape restrictions. After several missteps, consensus around a final scheme emerged in 1999. The design evolved from a scheme of 100-meter height to 180 meters that unexpectedly was proposed by the city authorities and their advisors arguing that a higher building would eliminate the "bulky" appearance of the design. In July 1999, Foster's solution was presented to the City Planning Authority's approval (Morrin, 2008).

Incidentally, these circumstances render this project a perfect candidate for DAV analysis.

DAV Analysis of the Swiss Re Tower

There was a confluence of requirements for this project from several fronts: client, architect, city planning authorities and their advisors. In the end, these requirements entailed ten categories (Table 10.2).

There were 11 stakeholders for the Swiss Re Tower. Planning authorities, authorities' advisers, and citizen organizations played an important role in the development of the final project design. The design process went through 21 months of negotiations and proposals dealing mostly with the height and shape of the building. Another important stakeholder group was the media that generated the "spin" and "buzz" that made the building well recognized among conservationists and the general public. All stakeholders were interested in the general impact of the building on economics, stature, productivity, and environmental quality.

Based on an analysis of the planning, decision-making, and execution of design of the Swiss Re Tower, we identified five potential VBD features: three-dimensional curved shape of the tower, engineering of a Diagrid steel structure, an environmentally sound cladding system, multistory light wells, and use of BIM in the documentation of the construction documents.

To demonstrate methods applicable to DAV analysis of Swiss Re, we will analyze two of its most impactful features in detail: building form and Diagrid structural system. These two features have the most impact on the building's design value.

Table 10.2 Final project requirements (Morrin 2008; Rossiter 2006)

1.	40 stories building, 1 basement, 2nd building for services
2.	Icon for City of London
3.	Baltic Exchange site
4.	360 degrees of views
5.	500,000 ft^2 office space
6.	200,000 ft^2 for lease—Higher rental space
7.	40–50% energy consumption—Class A performance specifications
8.	First sustainable building in London
9.	Awards—See Table 10.2
10.	Light wells—Interaction between floors

VBD Feature: Building Form

The development of Swiss Re Tower's building form is the result of a synthesis based on a number of criteria, which are a direct response to the particular site and client requirements.

The form was achieved in two ways: first, it was based on a radial plan with circular perimeter, and, second, its diameter varied for each floor. It widened as it rose from its base and then tapered towards the peak of the tower (Fig. 10.2). This allowed the floors at the middle of the tower to offer maximal floor area (Bridwell 2000). The radial plan of the tower responded to the constraints of the site, less space on the ground level and less of a footprint on the ground helped maximize external public space around the building (Fig. 10.3) (Shen 2009).

The diameter of the tower is reduced at street level which also opens up the areas around adjacent buildings. The tapering profile allows the reduction in floor area as one moves towards the top of the building, reaching a spatial climax at the glazed domed roof form that enhances, but does not dominate, the London skyline (Munro 2004). The shape also appears slenderer than a rectangular block of equivalent size and occludes less of the sky. This unusual form was preferred not only because of its unique appearance but also because of the following positive aspects:

- *Reduce amount of wind deflected to the ground:* Due to its aerodynamic tapering profile, there is a reduction in the amount of wind deflected to the ground as compared with rectilinear towers of similar size that helps maintain pedestrian comfort and safety.[1]
- *Minimize wind loads on the structure and cladding:* The building form encourages wind to flow around its face and minimize the wind loads on the structure and cladding.

[1] Fosterandpartners.com, "Awards by Project," 2010, last accessed October 2010. http://www.fosterandpartners.com/Data/Awards/S.aspx

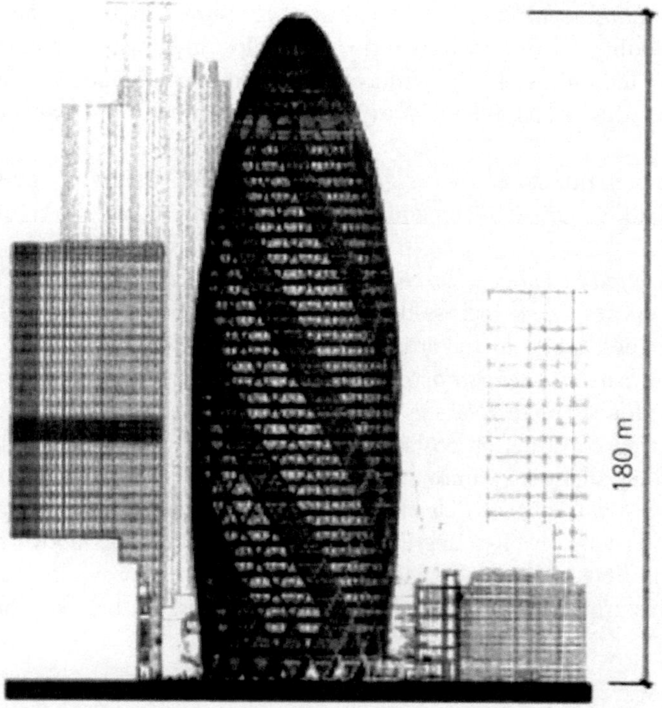

Fig. 10.2 Tapered form of Swiss Re

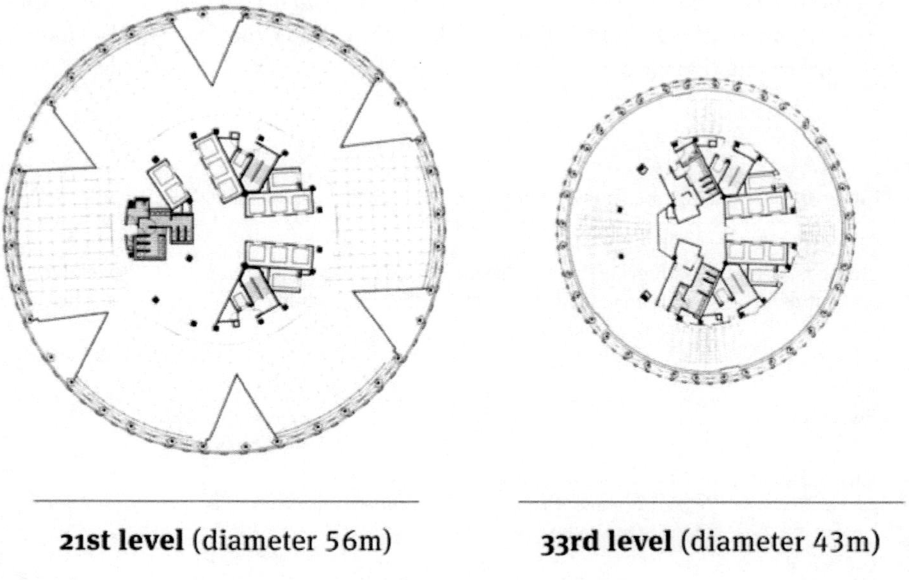

Fig. 10.3 Floor plans of Swiss Re (Powell 2006)

- *Assist natural ventilation:* External wind pressure differentials are generated by the building's aerodynamic form that helps in assisting natural ventilation through the light wells and reduces the requirement for conventional air conditioning. This reduced the cost of HVAC and lighting in the entire tower (Fedun 2005).
- *Maximize plaza space on the ground level.* Radial plan with circular perimeter had the maximum diameter and the space at the ground level where the plaza was placed.
- *Enhance public space at street level:* The plaza space created by the radial plan opens up new views across the site to the frontages of adjacent buildings and allows good access to and around Swiss Re's development.
- *Reduce heat loss and solar gain:* The circular building form of the Swiss Re Tower requires up to 25% less surface area than a rectilinear building. In addition to reduced construction costs, this means less heat loss in the winter and solar gain in the summer (Munro 2004).
- *Provide better physical and visual interconnectivity:* Each floor plate of the tower is rotated by five degrees with respect to the floors above and below (as interconnected levels of a light-well). In addition to providing natural ventilation, this provided better physical and visual access for the users on each floor.

Building Form: Cost–Benefit Analysis and Benchmarking

We estimated the cost (Table 10.3) and benefit (Table 10.4) values for the building's form with respect to its stakeholders—*Architect/Engineers*: Norman Fosters and Partners; *Client*: Swiss Reinsurance Company; *City and Country*: London, Britain; and *Culture and Media*: exhibitions, books, publications, movies—in both quantitative and qualitative terms.

Table 10.3 Cost assessments of building form for each stakeholder category (Powell 2006)

Stakeholders	Qualitative	Quantitative
Norman Fosters and Partners	Criticism about the shape, (not pleasing and does not fit in urban space around it)	Cost incurred and time spent during planning and feasibility stage (Aug 1996 to Jul 1999). Cost of conducting wind tunnel tests and aerodynamic studies
Swiss Re	–	Cost of circular slabs = £ 4,981,663.50
City of London	Criticism about the shape, (not pleasing and does not fit in urban space around it)	–
Culture and Media	–	–

Table 10.4 Benefit aspects of building form for each stakeholder category (Powell 2006; Lawrence and Samuel 2009)

Stakeholder	Qualitative	Quantitative
Norman Fosters and Partners	Awarded the Royal Institute of British Architects' (RIBA) ***Stirling Prize*** in October 2004. Identified as one of the ***seven next wonders of the world*** by US edition of *Conde* Nast Traveler, April 2005. ***Design***: Helps in maintaining pedestrian comfort and safety at the base of the building as wind is not deflected to ground level. Minimizes wind loads on structure and cladding. Improves wind conditions in the vicinity (as per wind tunnel tests)	***Maximized the area*** on the ground for public spaces (as compared to square plan geometry) ***13%*** increase (=4994 m^2). ***Maximizes natural light***: Reduction in heating, lighting and ventilation costs = ***£82,278.336 ($129,000)*** per year. Development of new software to cope with the complexities of the design (structural and parametric), reduction in construction cost: NPV ***$5,376,414.80*** (2–5%) refer aggregated evaluation section. Requires up to **25%** ***less surface area*** than a rectilinear building, less heat loss in winter or solar gain in summer
Swiss Re	***Rebranded Swiss Re***, the reinsurance business company	Revenue generated from renting the office space. ***Swiss Re Tower Rents***: Bottom floors: NPV = £65 ($102)/ft^2/annum; upper floors: NPV = £70.1 ($110)/ft^2/annum
City of London	***Rebranded the City of London*** and has become a symbol of London. Changed the ***perception of tall buildings*** in London. The ***building outline*** now appears in a variety of images such as logos, promotional posters, artwork, posters, and publicity materials about London	Gave rise to the ***approval*** of a large number of interesting high building projects in London. Price escalation of land in immediate vicinity
Culture and Media	***Recognition in exhibitions*** even before its completion, Great Expectations, at New York City's Grand Central Station in 2001. Intensive media coverage of the building from the time the design was unveiled. ***Appeared on the cover of magazines*** (Time Out's 2005 guide to London, Architecture Today, Architectural Review), newspapers (The Guardian) and journals (RIBA Journal, Italian design journal *Domus*, Wall Street Journal). Number of lectures, writings, books, articles, citations published on Swiss Re	Revenue generated by exhibitions, books, publications. Revenue generated from movies: Appeared as the setting in movies like Match Point, Basic Instinct II, and Bridget Jones. Revenue generated by documentary film "*Building the Gherkin*" (documentary of the design and construction process)

To measure and compare the form and shape of Swiss Re Tower with other buildings, we used a Benchmarking tool developed by the Construction Industry Council (CIC)—Design Quality Index (DQI). It is used to measure the quality of buildings. This Benchmarking tool includes space usage, construction, engineering, and sustainability measures. It identifies three aspects of design quality: *Functionality, Build Standards,* and *Impact on Users* subdivided into elements to allow quality comparisons.[2] In the DQI process, it is possible to compare generic forms of buildings, using the five-point scale previously introduced: i.e., high, medium-high, medium, medium-low, and low or 0.99, 0.75, 0.50, 0.25, and 0.01, respectively (Howard 2006).

Comparing the NPV of the cost and benefit values of the Swiss Re Tower's form, we observed that its benefits far out rank its costs. Its distinctive form not only rebranded the Swiss Reinsurance Company, but also the City of London. It became a landmark in its own right and also led to the approval of a large number of interesting high-rise projects in London. Its outline now appears in a variety of books, magazines, journals, and other publicity material referring to London.

The general implication of this finding is that the conventional, rectilinear form is the safest and most economical in most situations, but more complex forms may come close to distinguishing the building in the DQI Benchmark measure. On the other hand, under ordinary circumstances, as complexity increases, the risk factor also increases. However, it is fair to assume that the value of form in iconic buildings like Swiss Re can be of a different category altogether.

VBD Feature: Diagrid Structural System

The steel Diagrid is the triangulated curved perimeter structural framing system used in the construction of the Swiss Re Tower among other tall building designs. Apart from providing stability to the tower and vertical support for the slabs, it is the rigid frame of the structural system. It is a self-sufficient structural element with a built-in service core that can take a variety of loading conditions. It also replaces the structural functions of perimeter columns and freestanding interior columns.

The Diagrid creates a circular pattern made up of linear steel segments that are connected by steel nodes/junctions at the intersecting points between these sections, and at points where the geometry of the tower floor changes. The steel sections form two-story-high A-frames which are placed end to end to form a frame that reads as a pattern of diamonds.

The diameter of the A-frame columns varies from 508 mm on the lower floors to 273 mm towards the 36th and 37th floors. Each A-frame measures 9 m at its base. The diagonal struts consisting of steel tubes are connected by a horizontal tie 250 mm in diameter. 540 mm deep plate girders, the wide flanged radial floor beams, span 14 m between the Diagrid and the central core. Adjacent beams are arranged on plan at 10° so that the slab has a maximum span of 4.5 m at the facade.

[2] "Design Quality Indicator," 2010, http://dqi.org.uk, last accessed April 2021.

The floor is a composite deck slab creating a horizontal diaphragm that assists in distributing the wind loads.[3]

Each of the nodes are 2 m high, and contain three steel plates which are welded together at varying angles to address the changing geometry of the structure, thus avoiding the need to fabricate individual end details to suit each location. Steel subcontractor Victor Buyck-Hollandia JV (VB-H) developed the node design details, including loading combinations involving primary structural actions, local floor eccentricities, and cladding loads, to meet a number of defined performance criteria:

- Robustness in tying requirements.
- Movement and restraint in requirements between the Diagrid system and the floor slab.
- Erection tolerances and fit with the cladding geometry.

At every other floor in line with the nodes, there are perimeter hoops that encircle the building, counteracting horizontal spread of the structure and binding the whole structure together. The hoops also turn the Diagrid into a very stiff triangulated shell that provides excellent stability for the tower. The Diagrid construction happens at two contiguous stories at a time. However, specialized steel construction was required on the top three floors of the building (38th–40th floors) that contains the domed structure.

The dome is 22 m tall and has 8 horizontal hoops of diminishing diameter. The steelwork required that the dome to be partially fabricated in Austria. Its segments were first assembled at ground level and then the final dome was assembled at the top. Some of the steel used in the project was from the remains of the Baltic Exchange building that used to exist on the site. Parametric modeling was used to resolve the complex geometry of the Diagrid and to determine the sizes of the steel frame.

One of the main advantages of a Diagrid is the inherent ability of the members to resist compression and tension stresses at once, creating a diminished need for steel in the building. This structural system has been used in many other projects such as the Hong Kong and Shanghai Bank in Hong Kong, Greater London Authority in London, and Hearst Tower in New York (McCain 2009).

Diagrid Structure: Cost–Benefit Analysis

The amount of steel and details of all the members used in Swiss Re are summarized in Table 10.5. We estimated the benefit and cost picture of Swiss Re's Diagrid structure in quantitative and qualitative terms, from the vantage point of view of the constituents of its design (Table 10.6).

[3] "Architecture Steel Stahl Acier, Swiss Re London," 2010, European Convention for Constructional Steelwork, http://www.steelconstruct.com

Table 10.5 Steel used in the Swiss Re Tower

Description	Quantity
Total amount of steel used	8358 tons
Diagrid (29% of total steel)	2423 tons
Total number of primary steel pieces	8348
Total length	35 km
Diagrid column size (ground—Level 2)	508 mm, 40 mm thick
Diagrid column size (level 36–38)	273 mm, 12.5 mm thick
Number of nodes	360
Top dome—Rectangular hollow sections	110 × 150mm welded

Source: *Architect, Client/User, City and Region, Contractors* (Skanska Construction) and *Structural Engineer/Steel Fabricators* (ARUP, Victor Buyck Hollandia Joint Venture Ltd., and Waagner-Biro)

Table 10.6 Benefit picture of the Diagrid system (Munro 2004; Moon et al. 2007)

Stakeholders	Qualitative	Quantitative
Contractor	Rigid structural frame Core can be designed as an open plan steel structure. Foundation loads reduced. Window panes and cladding customized to the frame, use of plane glass. Interior columns are not required	Structural efficiency by adding to building stiffness by 40% An increase in area of 185 sq. m for a typical floor
Structural Engineer/Steel Fabricators	The structure allowed for cutting down use of steel by 20% Parametric modeling made process of fabrication and construction simpler and faster	Cost savings of £85,606 Construction cost savings of approx. £2.76 million (2% of total construction cost) by using BIM
Architect	Allowed for irregular design of floor plates. Helped realize the complex building form. Expertise in building the structure resulting in minimum errors	Time and cost savings
Client/User	Immense recognition for high profile design. Column-free interior spaces. Natural light and ventilation	Calculated in building form. Calculated for building form
Environment	Iconic design. Wide use of Diagrid in other structures	

For heights greater than 40 stories, the Diagrid structure is structurally superior compared to other structural types. The only disadvantage is material usage in the fabrication of its nodes. In terms of savings, Diagrid is suboptimal as opposed to braced tube structures; however, it saves between 2 to 20% of material against conventional braced tube steel structures, up to 80 stories (Tables 10.6 and 10.7).

Table 10.7 Advantages and disadvantages of the Diagrid systems (Ali and Moon 2007)

Type of structure	Material	Height limit	Advantages	Disadvantages
Framed Tube	Steel	80	Resists lateral loads	Shear lag hinders true tubular behavior
Braced Tube	Steel	100	Resists lateral shear, wider column spacing possible, reduced shear lag	Bracings obstruct view
Bundled Tube	Steel	110	Reduced shear lag	Interior planning limitations
Diagrid	Steel	100	Efficiently resists lateral shear	Complicated joints
Diagrid	Concrete	60		Expensive formwork and slow construction
Space Truss Structure	Steel or concrete	150	Efficiently resists lateral shear	May obstruct view
Super frame	Steel	160	Could produce super tall buildings	Building form depends on structural system
Exoskeleton	Steel	100	Interior floor never obstructed by perimeter columns	Thermal expansion and contraction

Table 10.8 Assessment of Diagrid material savings in comparison to Braced Tube (Moon et al. 2007)

Building height	Braced Tube Steel Mass	Diagrid Steel Mass	% of material usage b/w Braced Tube vs. Diagrid
40 stories	989	821	20.5%
50 stories	2128	1855	14.7%
60 stories	4113	3822	7.6%
70 stories	7222	6882	5.9%
80 stories	11,848	11,574	2.4%
90 stories	18,545	18,412	0.7%
100 stories	27,287	27,703	−1.5

Table 10.7 presents a comparison of the Diagrid system with steel or with concrete to Framed Tube, Braced Tube, Bundled Tube, Space Truss, Super frame, and Exoskeleton systems. A strong cost competitor to Diagrid system, Braced Tube Steel Mass system, is provided in Table 10.8.

Economic Analysis of the Swiss Re Tower

Exchange Value

In 1994, the Exchange Baltic Company sold the land and the old building to Trafalgar House at a price of £12 Million. This low price was due to the restriction over the site to restore historic structures. In 1998, Swiss Re compromised the

Table 10.9 Summary of the comparison of Swiss Re with other towers (Courtesy of Vinit Kumar Jain, Bhavna Muttreja, Alejandra Munoz Munoz)

Conversion: 1 m^2 = 10.7 ft^2	Rental price	Average price	Rental/average price ratio	Over price per sq. ft.	Office/site area ratio
Swiss Re	$10.78	$9.42	1.14	$1.40–$3.73	8.19
Agbar Tower	$3.97	$2.3	1.69	$1.77	0.6
Hearst Tower	$6.04	$5.75	1.04	$0.24	21.7

purchase of the site from Trafalgar House. Media reported a sale price in the range of £80–£100 Million (Powell 2006). This new high value of the site reflects the changes on the restrictions regarding the restoration of the damaged Baltic Exchange building. Subsequently, the Swiss Re was sold to the German property firm IVG Immobilien and UK investment firm Evans Randall at the price of £600 million.

The bottom floors of the Swiss Re Tower rents for £55 per sq. ft. per annum, while the upper floors, total of 50,000 ft^2, go for £60 per sq. ft. per annum. When fully occupied the total yield of the building ranges between £54 and £72 per sq. ft. per annum (Rossiter 2006). In London, the average office rental rate during the 2010s was £54–£72 per sq. ft. per annum[4] which breaks down into base rent £35–£45 per sq. ft. per annum, added rates £12–£18 per sq. ft. per annum, and services £7–£9 per sq. ft. per annum. Comparing the Swiss Re to other office towers we find that its exchange value is considerably higher (Table 10.9).

Use Value

The use value of the Swiss Re Tower has both negative and positive aspects. Negative aspects are the discomfort of tenants because of the lack of visual privacy of the flexible floor plans. As a consequence, partitions were added in some floors, interfering with the natural ventilation system. Presently, the only tenant that uses the natural ventilation system is the Swiss Re Company. Another adjustment made by the tenants of the building was the lowering of the temperature control from 26 °C to 24 °C, increasing the energy consumption for cooling.

Positive aspects of the Use Value of the Swiss Re Tower are the morale-boosting effect over the workers that the famous building provides. Over 25,000 have visited the restaurant and enjoy the 360 degree-view that the upper floors provide.

[4] Find a London Office, http://www.findalondonoffice.co.uk/toolbox/rental-guide, last accessed April 2021

Aggregated DAV Evaluation

Due to the innovative structural system of Swiss Re a savings of £85,606 was realized from material, labor, and fabrications processes. Given the inflation rate of 2.65 between 2001 and 2009,[5] an NPV savings of £105,529.97 or $166,758.46 was realized. Total construction cost savings due to innovations in the construction systems of the tower equal £2,760,000.[6] Given the inflation rate of 2.65 between 2001 and 2009,[7] this represents an NPV savings of £3,402,363.51 or $5,376,414.80.

Energy savings, a critical criterion in our analysis, is at a consumption rate of 568 kw per square meter per year, estimated based on the "Typical Prestige Air Conditioned Office Building." The *Good Practice* target consumption, based on this standard, is 348 kw/m^2/year. Swiss Re energy consumption turned out to be 215 kw/m$^{=}$/year, due to the cladding design and natural ventilation being used only 40% of the year. Using the aforementioned inflation correction this represents an NPV savings of £82,278.336 or $129,000.00/year.

On the other hand, there were substantial costs due to maintenance and operations functions, yet another important factor for building evaluation. Maintenance costs for Swiss Re equal £4,402,200 per annum (corrected for inflation = £5,039,270). There were added maintenance costs due to the special form and structure of the building: £251,963.50 or $398,152.72.[8] The total VBD calculation for Swiss Re based on its form and structural system is estimated based with the following assumptions:

- Net cost of finishing and fire protection of Diagrid could not be quantified.
- Revenues from use of the building's image in publications, movies, and other media have not been included.
- Additional costs and benefits resulting from features like glass cladding and maintenance, or sustainable design features have not been included.

With these caveats the aggregate cost–benefit value of the two VBD features of Swiss Re as analyzed here is equal to (Table 10.10): $5,341,010.54–$5,452,501.04.

[5] Williamson, S. H., "Seven Ways to Compute the Relative Value of a U.S. Dollar Amount, 1774 to present," Measuring Worth, 2009, last accessed April 2021. http://www.measuringworth.com/calculators/inflation/result.php

[6] Reinhardt, J., "Computer-Aided Construction Management," guest speaker at Course 12–611, Carnegie Mellon University, Pittsburgh, PA, November 30, 2010.

[7] Williamson, S. H., "Seven Ways to Compute the Relative Value of a U.S. Dollar Amount, 1774 to present," Measuring Worth, 2009, last accessed April 2021. http://www.measuringworth.com/calculators/inflation/result.php

[8] "Table 1—Importance of Maintenance and Operational Costs (Based on 2002–3 Typical Cost Figures by BMI)," Whole Life Costing, Consulting Excellence, 2004, http://www.constructingexcellence.org.uk, last accessed April 2021.

Table 10.10 Aggregate VBD cost–benefit picture for Swiss Re based on the form and Diagrid structure features (Courtesy of Vinit Kumar Jain, Bhavna Muttreja, Alejandra Munoz Munoz)

Category	Benefits–risks	Costs–risks
Energy savings	\$129,000.00	–
Structural system	\$166,758.46	–
Construction cost (parametric modeling)	\$5,376,414.80	–
Increase in rental value	\$66,990–\$178,480.50	–
Added maintenance costs (due to structure and form)	–	\$398,152.72
Totals	\$5,739,163.26–\$5,850,653.76	\$398,152.72
Net benefits (form and Diagrid structure)	\$5,341,010.54–\$5,452,501.04	–
Savings/m^2 (rentable area = 47,850/m^2)	(\$111.62–\$113.95)/m^2	–

Exercise 9

Are there any gaps in the DAV analysis of the Swiss Re Tower presented above? If your answer is affirmative, investigate these gaps and provide new data not included here and discuss their potential impact on the conclusions drawn.

Chapter 11
The Commerzbank Tower: Analysis of Another Seminal Case

Overview

Standing at a height of 300 m, the Commerzbank Tower (Fig. 11.1) was also the tallest skyscraper in Germany and held that title for all of Europe from its erection in 1997 until 2005 (Arand et al. 2013).The structure boasts a full-height atrium, nine four-story sky gardens, a Vierendeel truss structural system, and a double-skin facade (Table 11.1). Together these features create a high-quality indoor environment that is less impactful on the environment because of reduced energy consumption and emissions compared to conventional high-rise office buildings. It is still considered the world's first ecological office tower.

Commerzbank Tower commissioned the architectural firm of Foster and Partners to design its headquarters building. Norman Foster's focus was on "*combining an ambitious structural and aesthetic vision with a quest for natural ecological solutions and the well-being of the individual.*"[1]

Foster and Partners' initial proposal in 1991 (Table 11.2) had pinwheel/projecting wings and allowed all offices to be naturally lit, but large surface area on the facade resulting from this configuration made the design expensive. The structure was altered to have a central atrium to provide natural lighting for the internal spaces and reduce the surface area of the facade. Additionally, the triangular floor plan was oriented away from the old Commerzbank Tower. At this phase of design, the sky gardens were on the south side of building and they contained elevators. This design then was changed to have sky gardens on all sides to account for concerns about the unevenness of the structural loads. In this scheme, the Vierendeel truss system was configured to more evenly distribute the structural loads by moving all utilities and elevators to the corners of the footprint. As a result, the office tower was converted into a column-free space on every floor.

[1] Foster and Partners (2013) Commerzbank Headquarters. Architravel; Online Architecture Guide http://www.fosterandpartners.com/projects/commerzbank-headquarters/ Accessed Apr 2019.

Ö. Akın, *Design Added Value*, https://doi.org/10.1007/978-3-030-28860-0_11

Fig. 11.1 The Commerzbank Tower Neue Mainzer Str. 32, 60,311 Frankfurt am Main, Germany (Image credit: "Frankfurt, Kaiserplatz, Commerzbank Tower" by Polybert49 is licensed under CC BY-SA 2.0)

Located in the heart of the business district in Frankfurt, Germany, the tower's structural design and environmental features were ahead of its time. The resulting structure turned out to be the world's first ecological office tower. Foster and Partners focused their design on user health and indoor environmental quality. They brought nature into indoor spaces by including four-story-high sky gardens in their design.

Although the design came with a construction duration of 4 years and cost 600 million DM, or $300 million US dollars, the benefits accrued by clients, users, subcontractors, and the City of Frankfurt made the effort worthwhile. In conducting the

Table 11.1 Building facts of the Commerzbank Tower (Courtesy of Laleh Gharanjik, Yasha Mir, Jessica Rinn)

Commerzbank Tower	Information	Source
Type	Office sky scraper	Foster and Partners
Construction phase	1994–1997	Lambot et al.
Design phase	1991–1993	Lambot et al.
Gross area	85,500 m^2	Commerzbank
# of work spaces	2400	Foster and Partners
Height	259 m (excluding 41 m of antenna)	Commerzbank
Levels	53	Foster and Partners
Cost	$408 mil	Arend et al.
Gross volume	538,000 m^3	Commerzbank
Structural system	Steel Vierendeel truss structure	Commerzbank

Table 11.2 Stakeholders of the Commerzbank Tower (Courtesy of Laleh Gharanjik, Yasha Mir, Jessica Rinn)

Stakeholders	Company name
Owner	Commerzbank AG
Architects	Foster and Partners, (Sir Norman Foster)
Structural engineers	Ove Arup and Partners, Krebs and Kiefer
General contractor	Hochtief AG
Quantity surveyor	Davis Langdon and Everest
Mechanical and electrical	Roger Preston and Partners/ RP&K Sozietat GmbH/ Petterson and Ahrens
Landscape architect	Sommerland and Partners
Lighting engineer	Lichtdesign
Foundation engineers	Ingenieursozietat Katzenbach und Quick

DAV analysis of this tower, two design features were investigated: the double-skin facade and the Vierendeel truss system. Additionally, using post facto analysis of the structure highlights strategies that can be applied to pre facto evaluation of similar features in other DAV cases.

It is also noteworthy that this commission involved the direct or indirect participation of a myriad of stakeholders (Table 11.2).

DAV Analysis

Due to its unique architectural design, Commerzbank Tower has many physical features that make it superior to most other high-rise buildings in Europe:

- *Form and Shape*: Commerzbank Tower has a hollow floor plan because of its full-height atrium which allows in natural light as well as sight lines into the sky gardens from its interior work spaces.

- *Triangular floor plan:* The floor plan with its unique shape and curved corners manipulates wind moment and gives the building a distinguished form, which is non-tapering, unlike many other high rises around the world.
- *Sky gardens*: Commerzbank Tower has nine naturally ventilated, four-story-tall sky gardens which encompass three different ecologies: Asian, North American, and Mediterranean. Each one of these sky gardens is situated between two eight-story-tall office stacks and creates a communal space for employees. The sky gardens have evenly distributed daylight due to their high glazed external walls and reflective ceilings.
- *The hollow floor plan:* Commerzbank Tower's use of the hollow floor plan in combination with the four-story sky gardens allows all work spaces to use daylight and natural ventilation with an unblocked view to the outdoors. Full-height windows and shallow floor plates guarantee admission of ample natural light that replaces artificial lighting through the most of the day.

We will carry out value analysis for two of the Commerzbank Tower's most significant design features: the *double skin facade*, and the *steel Vierendeel truss structure system*.

Spin and Buzz[2]

The Commerzbank Tower gained immediate media attention during construction due to its innovative design. It was deemed to be the "*first ecological tower in the world*" and published in numerous architectural titles. Two books: one named *Sir Norman Foster and Partners: Commerzbank, Frankfurt am Main,* by Volker Fischer, and the other named *Commerzbank Frankfurt: prototype of an ecological high rise,* by Davies. C, Lambot, were published in *Tower Architecture* in 1997, right after Commerzbank's construction was completed.

The skyscraper has won numerous awards regarding its sustainable and green design, including the *Green Building Award* of the City of Frankfurt, *RIBA Architecture Award, Bund Deutscher Architekten, Martin-Elsaesser-Plakette Award,* and *British Construction Industry Award.*

In recent years, Commerzbank Tower has also appeared in other forms of media such as *Euro Contemporary tile set in SimCity 4, background of Motorola RAZR cell phone*, and *Covert Affairs episode*.

The timeline outlined in Table 11.3 depicts the significant events that helped shape the Commerzbank Tower design.

[2] Toker argues that the promotion of a design starts with "spin," which is putting out into the world the features that make a design noteworthy or even outstanding. Then, initiated by the spin, "buzz" takes over as a natural course of events native to public media. Then the feature in question becomes a legend.

Table 11.3 Timeline of Commerzbank Tower (Courtesy of Laleh Gharanjik, Yasha Mir, Jessica Rinn)

1960–1980	Commerzbank occupied 30 buildings and was looking to consolidate them.
1980s	The local government restricted development of new skyscrapers due to public outcry.
1989	Public dislike of skyscrapers had relaxed and a site became available for construction, provided that the design met government standards, such as benefiting the public in Frankfurt, addressing environmental concerns and including accommodations for international companies.
1991	In the month of February, twelve design companies including nine from Germany, two from the USA and one from the UK, were invited to provide submissions for the construction of the new Commerzbank Headquarters. The companies were given a deadline until June 3rd.
1992	By September, the cost estimation was completed and the project delivery method was selected.
1993	All subcontractors were hired.
1994	In August, construction detailing of the foundation began.
1995	By June, the pilings and the main foundation were completed.
1995	In October, the installation of the steel frame had reached the 29th floor.
1995	By December, the first cladding panels arrived on the construction site and the steel frame reached the 41st level.
1996	In March, the steel frame was completed and installation of building services started. The atrium and exterior facade cladding reached the 27th story.
1996	Nine months later, the 40 m long antenna was lifted into position.
1997	By March, installation of interior fixtures started and the sky gardens trees were delivered to the construction site.
1997	Commerzbank Tower was officially opened on July 24th.

Stakeholders

A variety of stakeholders were involved in this project. Their benefits accrued by these stakeholders including fame, income, function, and sustainability are as indicated in Table 11.4.

- *Client—Commerzbank*: Commerzbank is a leading bank in Germany, founded in 1870 with more than 56,000 employees and 10,000 billion Euros in annual revenue. The Commerzbank Tower has brought recognition to the client group. The structure has certainly impacted the company's users. As a result of the sustainability features, the energy consumption in the building has been dramatically reduced resulting in recognition of the Commerzbank as a leader in energy saving.
- *Architect*: Foster and Partners are renowned for their innovative and environmentally friendly architectural designs. This project has brought them greater recognition, since it is "*the world's first ecological office tower.*" In addition, it was recognized as the tallest skyscraper in Europe at the time of its construction.

Table 11.4 Stakeholder benefits of the of Commerzbank Tower (Courtesy of Laleh Gharanjik, Yasha Mir, Jessica Rinn)

Stakeholders	Fame	Income	Function	Sustainability
Client—Commerzbank	+	+	+	+
Architect—Foster and Partners	+	+	+	+
Structural engineer—Arup and Partners	+	+	−	−
Media	+	−	−	+
City of Frankfurt—Inhabitants	+	−	+	+
Authorities	−	−	+	+
Contractor	+	+	−	−
Facade fabricators (Josef Gartner and Co.)	+	+	−	+
Users (Commerzbank employees)	−	−	+	+

Foster and Partners also gained fame because of their focus on the building's sustainability status.

- *Structural Engineers*: Arup and Partners, a famous engineering firm with more than 10,000 projects, including various well-known architects including Jorn Utzon, on the Sydney Opera House (Chap. 16).
- *Media*: There has been significant "spin" and "buzz" of this building in the media especially right after its construction. There was intense media coverage on how such a large tower could be so sustainable.
- *City of Frankfurt's inhabitants*: Commerzbank Tower has been a landmark for Frankfort since its construction. It has brought significant attention upon Frankfurt as a progressive European City. Additionally, the public has benefited directly from having a building in their city consuming little energy and producing less emissions.
- *Authorities*: This building has brought fame to local government authorities who encouraged the construction of the skyscraper to take place. Due to the progressive features of the building, they were recognized as innovative implementers of policy and law.
- *Facade engineer (Josef Gartner and Co.*[3]*):* The facade of this tower is one of its most significant physical features that sets it apart from other high rises; thus it has distinguished the facade fabricator as an innovator of quality designs.
- *Users—bank employees*: The employees are invested in the design of the building because it is their place of work. The structure's ability to function as a sustainable building with natural ventilation, ample daylighting, and views to the outside benefited their productivity and positive mindset at work.

[3] Josef Gartner and Co. is currently a part of Permasteelisa Group.

DAV Feature Analysis of the Double-Skin Facade

The double-skin facade enclosing the Commerzbank Tower highlights its ecological design. Foster and Partners hired Josef Gartner and Co., one of the world's leading envelope designers and manufacturers, to develop a double-skin facade that would be able to function at all heights of the skyscraper and conserve energy throughout the year, even during seasonal changes. Unlike conventional office buildings which have single-skin facades and need large capacity HVAC systems, the Commerzbank Tower incorporates two skins into its facade to allow for passive ventilation, heating, and cooling. As a result, the envelope saves cost by reducing energy consumption.[4]

The exterior skin (Fig. 11.2) consists of disjointed sections of fixed glass. Each section is made of (1.4 m) (2.25 m) sheets of 8 mm thickness glass with a 20 cm gap between each layer (Poirazis 2004). While the exterior layer of glass is a wind barrier, the gap acts as an air channel that promotes natural ventilation throughout the space between the two layers.

The interior skin contains double-glazed operable panes (Fig. 11.3). In total, there are 3100 motorized interior panes that can be controlled either by the user or a monitoring system that is constantly adjusted to interior environmental conditions. Depending on the weather, the interior windows and air channels can be opened and closed. During mild weather, both facade layers allow temperature regulation through natural ventilation. On the other hand, during extreme weather conditions, i.e., cold winter or hot summer days, the panel system is closed to back-up the HVAC system.

Risks

Foster insisted on a building with high indoor environmental quality by regulating the natural environment rather than over reliance on mechanical systems. A double-skin facade for a skyscraper had never been constructed in Frankfurt before. To reduce the potential for failure of the facade system Gartner and Co. researched and developed a cladding structure capable of withstanding excessive winds, humidity, and thermal expansion. Using a $15 million investment, Gartner and Co. developed a model to predict the environmental performance of the Foster design. Based on this research, a passive double-skin facade system was estimated to be annually effective, 60% of the time.

Additional risks were due to the potential for glare since the transparent facade would surround all three sides of the triangular form. Even the interior walls, facing

[4] The reader is encouraged to compare this installation with the difficulties encountered in the construction of the double glazing system used in the John Hancock Tower in Boston, MA (Chap. 12).

Fig. 11.2 Exterior skin of facade

Fig. 11.3 Open windows of interior skin (Rendering by Yavuz Göncü)

the building perimeter that was made of glass to allow as much natural light into the office space as possible, further increased the risk of glare and its negative effects on the office environment. Constructing the facade also came with potential risks since the tower was being constructed in the heart of Frankfurt's business district. Since the construction site had to be compact and had limited storage space, the glass panels had to be delivered to the site on an as-needed basis. In order to avoid construction delays, the entire operation had to be closely managed.

Costs

A primary expenditure during the design process was the $15 million paid to Gartner and Co. to conduct their research and design the double-skin envelope system (Hartkopf 2013). This research funding was absorbed by Commerzbank as part of the architectural design fee. Without this initial investment, the Commerzbank Tower's performance would not have functioned as efficiently. Similar facades with operable panes can range from $275/m^2$ for a single-glazing system to $1200/m^2$ for a double-glazing system.

The building structure was estimated to cost $916/m^2$ with a total cost of $10.8 million during the time of construction. The double-skin facade initially cost much higher than single-glazing facades due to the use of roughly double the quantity of glass, the need for additional cladding materials, and labor-intensive conditions that were required. In addition, the double-skin facade system comprised four layers of glass in most of the cladding areas, compared to just two in a conventional high-rise facades.

Some of the details of the NPV estimates for 1997–2013 of the Commerzbank are as follows[5]:

- Floor to floor height = 3.75 m.
- Surface Area of facade = (24 + 34 + 36 office wings) (3.75 m) (34 m) =11,730 m^2.
- Material and Construction = (Cost per m^2) (Double-skin area) = ($915.83/m^2) (11,730m^2) = $10,742,742 (in 1997-$).
- Research Investment = $15 million (in 1997-$).
- Operating Costs = (7 €/m^2)*($1/(1.0919 €))(11,730m^2) = $75,200 per year (in 2003-$).
- Cleaning Costs = (3.9 €/m^2)* ($1/1.0919 €)(11,730m^2) = $41,900 per year (in 2003-$).
- Total NPV costs = (1997-costs) (1.03) + ··· + (2013-costs) (1.03) = −$43.9 million.
- The double-skin facade's NPV estimate is $43.9 million.

[5] A 3% inflation rate was assumed in these NPV calculations.

Increased maintenance costs associated with the mechanically operable windows and blinds further exasperated the cost of the cladding system. Mechanically moveable parts of the facade system would have required additional repairs. Commerzbank had to spend $75,200 on operating costs and $41,500 on cleaning costs.

The down side of using natural ventilation to regulate airflow in a high-rise office is the presence of extensive variable conditions in the indoor environment. The Commerzbank Tower is designed to naturally ventilate for 60% of the year, which means that as the seasons change, so do interior environmental conditions. Temperature, humidity, and lighting levels fluctuate throughout the year and even throughout the day. Users had to be willing to work, to a degree, in a volatile indoor environment due to the double-skin facade as complex as the one in this case.

Benefits

Gartner's envelope design was more energy efficient than initially predicted. Rather than using the double-skin facade for 60% of the year, the system was actually passively ventilating, heating, and cooling the building's office spaces during 80% of the year. Additionally, the users enjoyed the natural ventilation so much that they were using it at a wider range of temperature settings. Days that were previously determined to be too cold or too hot were now within the users' thermal comfort range. Ultimately, the double-skin facade saved Commerzbank $464,000 annually in energy consumption in comparison to conventional office buildings.

- Electricity Consumption for Conventional Office Buildings = 15.5 kWh/ft^2
- Floor area = 920,314 ft^2
- Ventilation = 5% of total electricity
- Space Heating and Cooling = 48% of total electricity
- Energy Savings = (80%)(53%) (15.5 kWh/ft^2) (920,314ft^2)=$464,000 per year
- NPV of energy savings (1997–2013) = (energy savings in 1997) (1.03) + ··· + (energy savings in 2013) (1.03) = $10.1 million.

The energy savings from the double-skin facade alone resulted in an NPV of $10.1 million, as estimated above, over a 15-year period, from 1997 to 2013. The employees still had all of the same benefits of natural ventilation and outdoor views, through the atrium spaces. In total, the NPV of the double-skin facade is $33.8 million as shown in Table 11.5, based on initial investment and subsequent returns broken down into each year, from 1997 to 2013.

This amount does not accurately represent the value of the design because it does not include the benefits accrued from increased rental fees, user productivity, and healthfulness.

Currently, Commerzbank occupies the entire office space available in the tower. In the future, the financial company could rent floor space out at a higher rate than other buildings in the Frankfurt area because of the beneficial aspects of the indoor environmental quality. The double-skin facade design has the potential to increase

Table 11.5 Net Present Value of Commerzbank's Double-Skin Facade (Courtesy of Laleh Gharanjik, Yasha Mir, Jessica Rinn)

Year	Initial investment	operating costs	Cleaning costs	Energy savings	Annual investment	Inflation factor	Present value
1997	$25,742,742	$75,200	$41,900	$464,000	-$25,395,842	$(1.03)^{-16}$	–$40,752,871
1998	$0	$75,200	$41,900	$464,000	$346,900	$(1.03)^{-15}$	$540,459
1999	$0	$75,200	$41,900	$464,000	$346,900	$(1.03)^{-14}$	$524,717
2000	$0	$75,200	$41,900	$464,000	$346,900	$(1.03)^{-13}$	$509,434
2001	$0	$75,200	$41,900	$464,000	$346,900	$(1.03)^{-12}$	$494,596
2002	$0	$75,200	$41,900	$464,000	$346,900	$(1.03)^{-11}$	$480,191
2003	$0	$75,200	$41,900	$464,000	$346,900	$(1.03)^{-10}$	$466,205
2004	$0	$75,200	$41,900	$464,000	$346,900	$(1.03)^{-9}$	$452,626
2005	$0	$75,200	$41,900	$464,000	$346,900	$(1.03)^{-8}$	$439,443
2006	$0	$75,200	$41,900	$464,000	$346,900	$(1.03)^{-7}$	$426,643
2007	$0	$75,200	$41,900	$464,000	$346,900	$(1.03)^{-6}$	$414,217
2008	$0	$75,200	$41,900	$464,000	$346,900	$(1.03)^{-5}$	$402,152
2009	$0	$75,200	$41,900	$464,000	$346,900	$(1.03)^{-4}$	$390,439
2010	$0	$75,200	$41,900	$464,000	$346,900	$(1.03)^{-3}$	$379,067
2011	$0	$75,200	$41,900	$464,000	$346,900	$(1.03)^{-2}$	$368,026
2012	$0	$75,200	$41,900	$464,000	$346,900	$(1.03)^{-1}$	$357,307
2013	$0	$75,200	$41,900	$464,000	$346,900	$(1.03)^{0}$	$346,900
						NPV	–$33,760,449

worker productivity and health in addition to energy savings. On average, employers in the private sector spend significantly more on employee salaries and healthcare benefits than on energy.

Benchmarking

To determine the value of the Commerzbank Tower's double-skin facade, its envelope was compared to three other buildings with similar facades and similar performance targets. These comparable structures are: Arag 2000 Tower, Torre de Cristal, and Bligh Skyscraper (Fig. 11.4).

Arag 2000 Tower was designed by Foster and Partners, in collaboration with RKW, Dusseldorf. It was the tallest building in Dusseldorf and served as the headquarters for the ARAG Group of the European Insurance Company. It has a double-skin facade with a shaft-box system and operable windows. The shaft is ventilated by louvers and the inner facade layer consists of low E-glazing vertical rotatable aluminum casements. Natural ventilation is available for 50–60% of the year; for the rest of the year, a high-performance HVAC system is used.

With its sculptural and angled form, Torre de Cristal is the tallest tower in Spain. It was designed by César Pelli and its shape mimics that of a crystal. The last floor of this office tower contains a vertical winter garden facing the West interior wall.

	Commerzbank Tower	**Arag 2000 Tower**	**Torre de Cristal**	**Bligh Skyscraper**
Location	Frankfurt, Germany	Dusseldorf, Germany	Madrid, Spain	Sydney, Australia
Completion year	1997	2000	2009	2011
Height (stories)	300m (53)	125m (31)	250m (45)	139m (30)
Area	85,500 m^2	38,100 m^2	77,000 m^2	42,700 m^2
Total cost	$300 mil	$63mil	$340 mil	$270 mil
Image				

Fig. 11.4 Benchmarking four comparable buildings (Image credit left: "Frankfurt, Kaiserplatz, Commerzbank Tower" by Polybert49 is licensed under CC BY-SA 2.0; Image credit center left: "ARAG-Tower" by ARAG Allgemeine Rechtsschutz-Versicherungs-AG is licensed under CC BY-SA 3.0; Image credit right: "No.1 Bligh Street_Sydney" by bobarcpics is licensed under CC BY 2.0)

It is a class A energy certified building and consumes one-third less energy than a similar office building in Madrid, Spain. All work places have daylighting and a direct view to outdoors. Additionally, it has a naturally ventilated curtain wall facade with automatic blinds that adjust the light by a central computer.

The Bligh Skyscraper, designed by Ingenhoven Overdiek Kahlen and Partners, is one of Sydney's most sustainable office towers. It is the first tower in Australia with a double-skin facade and automatic external louvers. The Bligh has solar panels on the roof and recycles 90% of its wastewater through its basement sewage plant. A naturally ventilated, full-height atrium space is designed in order to provide ample daylighting for all work spaces. Also, the double-skin facade is naturally ventilated and decreases solar heat gain while promoting exterior views and daylighting.

Benchmarking

A modified form of the Analytical Hierarchical Process (AHP) is used to analyze the benchmarking of the four buildings' double-skin facades.

AHP is a systematic method used to compare multiple alternative features. A set of criteria is developed and given weights of importance. In a pairwise manner each option is compared for each criterion. The pairwise scale measuring the comparison between each building's feature against others ranges from 1 to 9, where 1 is *equal*, 3 is *moderate*, 5 is *strong*, 7 is *very strong,* and 9 is *extreme*.

These values signify how a facade compared to another according to the specified criterion. The evaluation criteria for the double-skin facade consist of:

- *Aesthetics*—the visual appeal of the exterior.
- *Performance*—the energy efficiency of the system.
- *Impact on the user*—the proportion of the facade dedicated to the quality of user work spaces or users' benefit attributed directly to the facade design.

We will use the AHP method to evaluate the benchmarking comparisons of the Commerzbank Tower against the three other high-rise office buildings along two feature categories: cladding and structure.

AHP Comparison of High-Rise Cladding

The weighted scores on the far right of Table 11.6 are normalized values, i.e., their sum to equal one, that show how well the building met the criterion in comparison to each other case.

Aesthetically, the Commerzbank Tower was found to be the best, followed by the Torre de Cristal, Bligh Skyscraper, and the Arag 2000 Tower (Table 11.6).

The double-skin facade around the Commerzbank Tower is separated into protruding sections that add definition to the building, whereas the other three buildings' facade lie uniformly along the sides of the building. The facade on the Torre de Cristal follows the twisting form of the structure, creating the illusion of a giant crystal structure. On the Bligh Skyscraper, the facade follows the interesting curvature of the building, whereas the Arag Tower's facade is monotonous and lacked unique aesthetic features.

Table 11.6 AHP comparison of double-skin facade aesthetics (Courtesy of Laleh Gharanjik, Yasha Mir, Jessica Rinn)

Aesthetics	Commerzbank Tower	Arag 2000	Torre de Cristal	Bligh Skyscraper	Weighted score
Commerzbank Tower	1	7	2	5	0.54
Arag 2000 Tower	1/7	1	1/5	1/3	0.06
Torre de Cristal	1/2	5	1	4	0.29
Bligh skyscraper	1/5	3	1/4	1	0.10
	1.84	16	3.45	10	~1.00

Table 11.7 AHP Comparison of Double-Skin facade for Performance Criterion (Courtesy of Laleh Gharanjik, Yasha Mir, Jessica Rinn)

Performance	Commerzbank	Arag 2000	Torre de Cristal	Bligh Skyscraper	Weighted score
Commerzbank Tower	1	4	1	1/3	0.19
Arag 2000 Tower	1/4	1	4	1/7	0.08
Torre de Cristal	1	1/4	1	1/3	0.11
Bligh skyscraper	3	7	3	1	0.55
	5.25	12.25	9	1.81	~1.00

Table 11.8 AHP comparison of double-skin facade's impact on its occupants (Courtesy of Laleh Gharanjik, Yasha Mir, Jessica Rinn)

Impact on occupants	Commerzbank	Arag 2000	Torre de Cristal	Bligh Skyscraper	Weighted score
Commerzbank Tower	1	1/5	1/5	1	0.08
Arag 2000	5	1	1/2	6	0.30
Torre de Cristal	5	2	1	7	0.54
Bligh skyscraper	1	1/6	1/7	1	0.07
	12	3.37	1.84	15	~1.00

Table 11.9 Summary AHP comparison of double-skin facade for all three criteria (Courtesy of Laleh Gharanjik, Yasha Mir, Jessica Rinn)

Torre de Cristal	= 0.29 + 0.11 + 0.54	= 0.94
Commerzbank Tower	= 0.54 + 0.19 + 0.08	= 0.81
Bligh skyscraper	= 0.10 + 0.55 + 0.07	= 0.72
Arag 2000 Tower	= 0.06 + 0.08 + 0.30	= 0.44

Performance scores of the four buildings were determined based on energy efficiency and emissions data (Table 11.7). Both the Torre de Cristal and the Commerzbank Tower use 1/3 less energy than conventional office buildings. The Commerzbank Tower benefited from its double-skin facade, saving HVAC system use 80% of the year. The Arag Tower utilized its facade for 50–60% annually. The Bligh's facade is the most efficient producing 42% less carbon dioxide than a conventional building.

Floor plans were analyzed to determine the impact of double-skin facade on its users (Table 11.8). Torre de Cristal has the highest impact on its users followed by Arag 2000. The Commerzbank Tower and the Bligh skyscraper have similar low-impact scores ranging from 07% to 08% of users directly benefiting from their double-skin facades.

After adding up the weighted scores of each building's Double-Skin Facade, the Commerzbank Tower was second best with a score of 0.81 (Table 11.9).

DAV Feature: Steel Vierendeel Truss Structure System

One of the primary features of Commerzbank Tower is its height of 53 stories, reaching almost 300 m above ground level. It was the first skyscraper in Germany to incorporate steel as its main structural component. Arup and Partners, the main structural engineers of the building, collaborated with a local German firm, Krebs and Kiefer, that advised them on local conditions.

The structure of Commerzbank Headquarters is a perforated tube structure which consists of approximately 18,000 tons of steel (Lambot 1997).The structure resists both the gravitational and the lateral forces without obstructing internal spaces with columns. It has three main components which work together to form its shape and support the perforations in its form, including to the spiraling sky-garden feature.

These components include the three *pillars* that are located at the corners of the triangular floor plan. Each pillar is made up of two H-section vertical steel component connected by beams and diagonal bracing, encased in reinforced concrete. These columns carry the entire load of the building and transfer it onto the secondary structural components: 111 telescoping pile foundation members located at the three corners of the triangular floor plan. The foundation members bear on the substrata soil that is mostly clay.

The third main component of the structure and the main DAV analysis feature is the Vierendeel truss system. The building is broken into four sets of eight-story Vierendeel trusses. These trusses are made up of eight horizontal members that are located on each floor level and four vertical members placed in response to the analysis of the wind effects on the structure. The Vierendeel truss structure distributes the weight of the building to the three corners of the triangle. The amalgamation of the structure in this manner and the use of steel have not only enabled the tower to reach a total height of 300 m, but also gave it a non-tapering form unlike most high rises (Fig. 11.5).

Risks

Commerzbank Tower was the first building in Germany to incorporate steel as its main structural component. This in itself was a risk, which involved expertise that was locally uncommon and a potential impediment to impact the cost and duration of construction. Additional risks could be attributed to the proximity of adjacent buildings and clay subsoil that was prone to excessive settlement. Without proper precautions, this would have led to existing buildings to lean towards the tower. In their initial proposal, Foster and Partners assumed a deep basement, accommodating a car-park and the main services for the building. After realizing potential foundation problems, they changed the structure from concrete to steel and eliminated

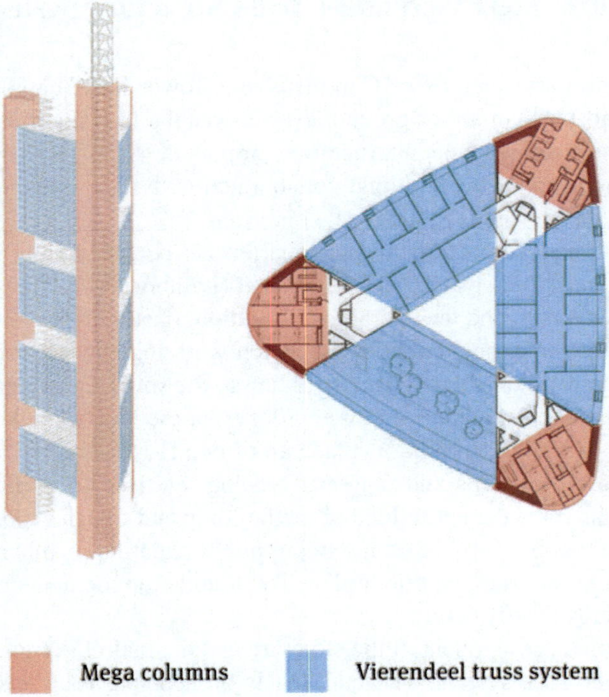

Fig. 11.5 Commerzbank plan showing location of Vierendeel truss structure and mega columns (Illustration by Yavuz Göncü)

the basement level all together. The veracity of these predictions was validated only after the excavation for the building commenced.

The switch to the use of steel, in lieu of concrete, required additional fireproofing expenses, especially for spaces adjacent to the steel members. This added cost was in the order of $10 million.

Costs

Although the total cost of the building was $300 million (Arend et al. 2013) and the total amount of steel used in the construction was approximately 18,000 tons, the cost–benefit analysis of the Vierendeel truss structure determined the value of the total NPV to be $34.16 million, as calculated below.

- Size of Horizontal ISMB = (475 mm) (1100 mm) = 0.13325 m^2 cross-sectional area.
- Total no. of floors = 53.
- Length of Horizontal Vierendeel truss system at each floor level = 34 m.

- Number of facade sides for the Vierendeel truss system = 3.
- Total Horizontal length of Vierendeel truss system = (53) (34) (3) = 5406 m.
- Total volume of Horizontal member ISMB = (cross-sectional area) (length) = (0.13325) (5406) = 720.35 m^3.
- Size of Vertical ISMB = (475 mm) (1000 mm) = 0.16575 m^2 cross-sectional area.
- Height of each floor = 3.75 m = 3 m with no members overlapping.
- No. of vertical Vierendeel truss system members at each floor level = 4.
- No. of facade sides with the Vierendeel truss system = 3.
- Total Vertical length of Vierendeel truss system = (3)(4)(53)(3) = 1908 m.
- Total Volume of vertical member ISMB = (cross-sectional area) (total length) = (0.16575) (1908) = 316.25 m^3.
- Total volume of steel used to construct the entire Vierendeel truss system = 720.35 + 316.25 = 1036.6 m^3.
- Density of mild steel = 7850 kg/m^3.
- Total density of steel used for the Vierendeel truss system = (7850)(1036.6) = 8140 metric tons – 1 metric ton = 1.1 American tons.
- Total amount of steel used in Vierendeel truss system = (8140) (1.1 = 9000 tons).
- Over 15 stories (including material cost + labor cost + equipment cost of ISMB per ton) = $3500.
- Total cost of using Vierendeel truss system = (1) × (2) = (9000) (3500) = $31.5 million.
- NPV = (31,500,000)(1 + 0.0274) (2013–2010) = $ 34.16 million.

However, in the absence of values estimated for maintenance costs and revenue generated based on the stature of the building, the total NPV for the steel Vierendeel truss structure system could not be accurately calculated. To accomplish this, the area differential due to the Vierendeel truss system compared to an ordinary steel structure needs to be estimated and this estimate would be multiplied with the average rental fee to obtain an accurate final estimate.

Benefits

The use of steel Vierendeel truss system was a major factor in making Commerzbank the tallest building in Germany and at the time of its completion, the tallest in Europe, especially considering the drawbacks of the soil conditions and the close proximity of neighboring buildings. The use of the steel Vierendeel truss system also allowed for a non-tapering form which not only increased the amount of rentable space, compared to other high rises. It also allowed for a 25% more glazing area than a concrete Vierendeel system. The use of concrete would have increased each beam size at the floor level to almost 2100 mm which is twice its current size (Arend et al. 2013).

Due to the non-tapering form, the structural system could be pre-fabricated and brought to the site. This benefited almost all of the stakeholders of the project. The structural engineers had to design only a limited number of construction details.

Prefabrication reduced assembly time of the structure by almost 20%. Since the use of the Vierendeel truss system reduced the total weight of the Commerzbank to approximately 132,000 tons, almost the same weight as the old 32-story existing concrete Commerzbank building right next to it, the possibility of the neighboring buildings to cave in due to weak soil conditions was all but eliminated.

The use of the steel Vierendeel truss system also allowed 400 m^2 of column-free space compared to a concrete construction. This had a positive impact on the office spaces as well the sky gardens. Within the office area, employees had unhindered views of both the sky garden and the city-scape. Additionally, the column-free space promoted easily reconfigurable functional spaces.

At the beginning of the buildings' use, Commerzbank Tower accommodated 2100 employees, while today it has 2400, a net increase of 300 employees. According to a post-occupancy case study, Fitz-Enz identified a reduction in turnover costs of $4100 per person, based on an average annual American employee salary of $45,000. This amounts to a significant benefit accumulated over many years of the tower's existence. Today, the Commerzbank occupies all of the office space available in the tower. Since the building boasts vast column-free spaces with superior daylight and views, it can attract significantly more rent revenue.

	Commerzbank Tower	**Silberturm Tower**	**John Hancock Center**	**Hearst Tower**
Location	Frankfurt, Germany	Frankfurt, Germany	Chicago, USA	New York, USA
Completion year	1997	1978	1969	2006
Height (stories)	300m (53)	166m (32)	457m (100)	182m (46)
Area	85,500 m^2	61,000 m^2	260,000 m^2	80,000 m^2
Total cost	$300 mil	$200mil	$100 mil	$500 mil
Image				

Fig. 11.6 Benchmarking Buildings with Advanced Structural Systems (Image credit left: "Frankfurt, Kaiserplatz, Commerzbank Tower" by Polybert49 is licensed under CC BY-SA 2.0; Image credit center left: "Silberturm @ Frankfurt" by *_* is licensed under CC BY 2.0; Image credit center right: "Chicago: John Hancock Center" by *rboed* is licensed under CC BY 2.0; Image credit right: "The Hearst Tower in Manhattan, New York" by o palsson is licensed under CC BY 2.0)

Benchmarking

In order to demonstrate the value of the Commerzbank Tower's steel structure, we compare its structural system to three other buildings with different structural systems. This includes Silberturm Tower, John Hancock Center, and Hearst Tower (Fig. 11.6).

Silberturm Tower: Designed by ABB Architekten and owned by Deutsche Bahn, Silberturm had been the tallest building in Germany until 1990. Its structural engineer is BGS Ingenieursozietät and it was founded on a 3400 m^2, 4 m thick reinforced concrete mat-foundation. Its structural system is concrete, which considerably constricts its height.

John Hancock Center: With its exposed structure on its facade, John Hancock Center is one of the most renowned buildings in Chicago, due to its structure's Expressionistic Style. It is designed by Skidmore, Owings, and Merrill. Because of the X-bracing system, it has column-free floors, which allows the building to be 457 m tall. It accommodates offices, residential apartments, restaurants, health clubs, and a pool. It has five enormous steel crosses on each side which support the beams. The steel used in this building is half as much as that which would be needed to build it with a conventional post and beam structural system. It weighs 384 million pounds and 46,000 tons of steel were used to construct it. Additionally, it has a tapering form—40,000 square feet at the base, 18,000 square feet at the summit—for increased stability and resistance to wind loads.

Hearst Tower: Known as the first "green" high-rise office building in New York City, Hearst Tower is designed by Foster and PartnersThe building, whose structural engineer is Cantor Seinuk, has a triangular, diagrid, framing pattern which requires 9500 metric tons of structural steel. It has pillars below the 10th floor that support its unbraced height. Using a diagrid system resulted in the consumption of 21% less steel than a traditional steel structure.

AHP Comparison of High-Rise Structure

The weighted criteria were used to compare the four buildings along three criteria.

- *Aesthetics*—the visual appeal of the shape and form, achieved primarily based on the structural system incorporated (tapering form).
- *Performance*—the time, height, and material efficiency of the system.
- *Ease of Construction*—the technology and labor needed for constructing the structural system.

Table 11.10 contains the matrix values for the AHP used to compare the structural systems of the four towers in terms of their aesthetic features.

The weighted scores on the far right are normalized values, i.e., their sum equals 1.00 that signifies how well the buildings meet each criterion. Aesthetically, the

Hearst Tower was found to be the best, followed by the Commerzbank Tower, John Hancock center, and Silberturm Tower.

Using diagrid in Hearst Tower gives it a unique and significant appearance due to its triangular framing pattern. Commerzbank Tower's mega columns and exposed X-bracing system in John Hancock center give them an exclusive visual presence. The Silberturm Tower's form and shape lack unique attributes.

The performance scores are determined by the time and material consumption needed to construct and maintain these systems (Table 11.11).

Table 11.10 AHP comparison of high-rise structures—aesthetics (Courtesy of Laleh Gharanjik, Yasha Mir, Jessica Rinn)

Aesthetics	Commerzbank	Hancock Center	Hearst Tower	Silberturm	Weighted score
Commerzbank Tower	1	5	1/3	6	0.23
John Hancock Center	1/5	1	1/5	3	0.09
Hearst Tower	3	5	1	7	0.60
Silberturm	1/6	1/3	1/7	1	0.06
	4.37	11.33	1.68	1.70	~ 1.00

Table 11.11 AHP comparison of high-rise structures—performance (Courtesy of Laleh Gharanjik, Yasha Mir, Jessica Rinn)

Performance	Commerzbank	Hancock Center	Hearst Tower	Silberturm	Weighted score
Commerzbank	1	3	5	7	0.60
Hancock center	1/3	1	3	5	0.22
Hearst tower	1/5	1/3	1	4	0.11
Silberturm tower	1/7	1/5	1/4	1	0.06
	1.68	4.53	9.25	17	~ 1.00

Table 11.12 AHP Comparison of High-Rise Structures—Ease of Construction (Courtesy of Laleh Gharanjik, Yasha Mir, Jessica Rinn)

Ease of construction	Commerzbank	Hancock Center	Hearst Tower	Silberturm	Weighted score
Commerzbank	1	4	4	8	0.62
Hancock center	1/4	1	1	5	0.16
Hearst tower	1/4	1	1	5	0.16
Silberturm	1/8	1/5	1/5	1	0.05
	1.63	6.20	6.20	19	~ 1.00

Table 11.13 AHP Comparison of High-Rise Structures (Courtesy of Laleh Gharanjik, Yasha Mir, Jessica Rinn)

Commerzbank Tower	= 0.23 + 0.60 + 0.62	= 1.45
Hearst Tower	= 0.60 + 0.11 + 0.16	= 0.87
John Hancock Center	= 0.09 + 0.22 + 0.16	= 0.47
Silberturm	= 0.06 + 0.06 + 0.05	= 0.17

Commerzbank Tower's structural system uses the steel Vierendeel truss system and mega columns, which gives it the highest score among the benchmarked buildings. The John Hancock Center's and Hearst Tower's structural systems use less steel than a traditional structural system does and the Silberturm Tower's concrete structure is not as efficient as the other benchmarked high rises. The steel Vierendeel truss system due to its repetitive pattern is the best building system among the four cases in terms of ease of construction, and Silberturm Tower's concrete structure was the worse (Table 11.12).

After adding up the weighted scores for each building, the Commerzbank Tower's steel Vierendeel truss system reached the highest score of 1.45 (Table 11.13).

Impact Analysis

Commerzbank Tower's impact on the stakeholders, the client, the architect, the structural engineers, the facade fabricators, and its users, was the highest, on account of the benefits provided by its double-skin facade and steel structure include:

Client—Commerzbank: At the first order which is the construction phase, construction of both double-skin facade and steel Vierendeel truss system had negative impact on the client due to costs. However, as time passed (second- and third-order effects) clients benefited from these features, due to less energy consumption (double-skin facade) and more usable space for work, which was possible due to less space taken up by columns on each floor and non-tapering form of the tower, both outcomes of the tower's structural system.

Architect—Foster: Construction of Commerzbank had positive impact on architect from very early on, when they won the design competition. The positive impact of the project on the architect increases at the second- and third-order effects because of the recognition that Commerzbank Tower brought.

Structural Engineer—Arup and Partners: Using the steel Vierendeel truss system had a positive benefit to structural engineers starting at the design and construc-

Table 11.14 Impact analysis of the Commerzbank Tower (Courtesy of Laleh Gharanjik, Yasha Mir, Jessica Rinn)

Stakeholders	1st order (construction phase)	2nd order (0–3 years)	3rd order (3–10 years)
Client—Commerzbank	–	+	+++
Architect (Foster and Partners)	+	++	+++
Structural engineer (Arup and Partners)	+	+	++
Facade fabricators (Josef and Partners)	++	+	0
Users (Commerzbank employees)	0	+++	+++

tion stages due to the income they earned. At the second- and third-order stages this positive impact increased because of the recognition that Commerzbank Tower brought to all stakeholders.

Facade fabricators (Josef Gartner and Co.): Building the high-performance double-skin facade brought the fabricators financial benefits during the construction phase.

Users—bank employees: Working in a sustainable work place brings about productivity and well-being. Natural ventilation and ample day light (which are the results of the double-skin facade) had a positive impact on the employees, after construction was been completed.

Generally speaking, as time passed, second- and third-order positive impact of the Commerzbank Tower on different stakeholders increased (Table 11.14).

Summary

Successful execution of DAV parameters, including features and functions, starts with good collaboration and communication between architects and engineers. It is possibly the only way to overcome the various hindrances in maintaining the design features from the initial design phase to the very end of construction. In the case of Commerzbank, several important goals were achieved:

- Commerzbank Tower served as a catalyst for sustainable development in Frankfurt.
- Double-skin facade is a unique design feature. Its driving force allowed the architect to incorporate it in terms of energy consumption and employees' quality of life.
- The driving force for an uncommon choice of structural system, the steel Vierendeel truss structure system, was a non-obtrusive, faster to construct form,

capable of supporting a high-rise structure and sky gardens without hindering the buildings' context.

- Additional physical building features, such as the unique form and sky gardens, are equally important as DAV features.

The double-skin facade was a risky design choice, but its impact on the users, energy savings, and its ability to emphasize Foster and Partners' vision of an ecologic l tower made this design option worth the effort. Such a facade had never been installed in a high-rise building in the region. In addition to the $10.7 million in materials and labor for the facade, Foster invested $15 million in research prior to any construction in order to further increase confidence in the design.

As a result of the highly efficient double-skin facade, the building does not require HVAC use for 80% of the time per year. This brings $464,000 in energy savings to Commerzbank. However, the most significant benefit is not in the energy savings, but the envelope's ability to create a high-quality indoor environment. Such an environment is conducive to its users' health because of its natural ventilation, natural lighting, seated access to nature, and user-controlled environment capabilities.

Although the NPV of the design option is $33.8 million, this single number is not an accurate assessment of the value the double-skin facade has for the Commerzbank Tower. In comparison to similar facades around the world, Commerzbank's double-skin facade fares better than the majority of tall building cladding systems in terms of aesthetics, performance, and impact on the user.

Commerzbank Tower would not have achieved the recognition that it did had it not been the tallest high rise in the country and at the time of completion, the tallest in all of Europe. The steel Vierendeel truss system played a pivotal role in that. It is also the first building in Germany to have used steel as a structural system, which introduced the entire country to a more contemporary and efficient means of construction in terms of construction time, assembly technique, and design possibilities.

Exercise 10

Are there any gaps in the DAV analysis of the Commerzbank Tower presented above? If your answer is affirmative, investigate these gaps and provide new data not included here and discuss their potential impact on the conclusions drawn.

Part IV
Selected Case Studies

This part contains six chapters, each dedicated to a case study of a landmark building suitable for DAV analysis. Readers are encouraged to conduct their own detailed DAV analyses of these case studies in Chaps. 12 through 17 that present preliminary data about John Hancock Tower in Boston; K. C. Hyatt Regency in Kansas City; Pruitt-Igoe in St. Louis, MO; Crystal Palace in London; Sydney Opera House in Sydney; and Citicorp Tower in New York, using the templates provided in the Appendix.

The chapters in this part contain a treasure throve of six building case studies:

Chapter 12	John Hancock Tower, Boston, MA, USA
Chapter 13	Kansas City Hyatt Regency , Kansas City, MO, USA
Chapter 14	Pruitt-Igoe, St. Louis, MO, USA
Chapter 15	Crystal Palace, London, UK
Chapter 16	Sydney Opera House, Sydney, Australia
Chapter 17	Citicorp Tower, New York City, NY, USA

Chapter 12
John Hancock Tower, Boston, MA, USA

"*A high wind blew in Boston on the night of January 20, 1993*" writes Robert Campbell, who has published a Pulitzer Prize winning expose on the travails and triumphs of the John Hancock Tower. This building is clearly the most celebrated building of the 70s both due to its successes and due to its failures. What happened during January of 1973 was the braking of dozens of the 4.5' × 11.5′ plate glass windows that were a part of Hancock Tower's cladding system, showering dangerous shards onto the sidewalks below (Schwartz 2001).

The building, arguably a masterpiece in terms of formal composition, strove to realize many innovations and take some design risks along the way while achieving its place in the history of Modern Architecture (Figs. 12.1 and 12.2). Among its innovations one needs to cite its massing, its reflective skin, and its *tuned mass damper* to keep it from swaying in the wind excessively. Yet, these features did not prevent the building's design from being flawed; some would argue this is the case because of some of these innovations rather than in spite of them. There were four major problems:

- Settlement causing structural damage in the adjoining Trinity Church by H. H. Richardson (Fig. 12.1) and underground utilities below Copley Plaza.
- Risk of excessive and structurally risky oscillations along the short direction of the building's foot print (Fig. 12.2).
- Risk of toppling of the building in the long direction of the building's foot print.
- Breakage of the plate glass window units (Fig. 12.3).

Eventually all of these problems were resolved. Today, having received numerous design awards, the tower stands as a safe and distinguished neighbor of H. H. Richardson's masterpiece, on Copley Square.

Ö. Akın, *Design Added Value*, https://doi.org/10.1007/978-3-030-28860-0_12

Fig. 12.1 H. H. Richardson's Trinity Church reflecting from the Hancock Tower glass window units

Fig. 12.2 The Hancock Tower, Copley Plaza (Image credit: "John Hancock Tower" by Peter Alfred Hess is licensed under CC BY 2.0)

Fig. 12.3 Plywood covered Hancock Tower (Image credit: © Spencer Grant)

Background

The siting of John Hancock Tower has been a difficult one from the beginning. In terms of urban and historical appropriateness it has attracted considerable attention since very early on. In 1968, during a conference organized at the Boston Architectural Center expressly for this purpose, it was hailed as an example of public and private decision-making in the City of Boston.

In the early seventies, during the foundation stages of its construction, adjoining buildings including the Trinity Church and underground utility lines sustained significant structural damage due to excessive settlement caused by Hancock's excavation cavity. The sheet piling and its lateral braces were inadequate.

Engineers monitoring the rate of excavation and settlement in adjoining structures observed almost perfect correlation with the rate of excavation. This problem was resolved after extensive repairs in neighboring buildings and installation of additional excavation bracing.

The remaining three problems were not manifested until one of them, namely the plate glass breakage started during a windstorm. Naturally, the attention of the engineers involved from the beginning, as well as those retained subsequently, was focused on lateral resistance and motion-related problems. Many diagnostic activities were undertaken based on data gathered from the site using sensors and laboratory tests.

Hanson Holley and Biggs, a Boston firm, was involved in solving the glass breakage problem. At first, they decided to study the building frame and settlement problems. Their first discovery was unrelated to the breakage problem. The building tended to sway excessively in the wind, which they concluded could be uncomfortable to the occupants. This was a state-of-the-art issue as not very many tall towers were being built at the time. John Reed, a PhD student at MIT, was doing pioneering work on comfort conditions in high-rise towers and was invited to join the team conducting analyses on the tower as well.

Soon it was decided that a Tuned Mass Damper system, like the one designed by LeMessurier for the Citicorp Building in New York City, would be installed at the top floors of the tower. Additionally, Mr. Hanson found in wind tunnel tests that the tower was subject to excessive oscillation due to combined effects of lateral and torsional forces in the short direction of the building's footprint. This was resolved by strengthening the bracing of the frame in the short direction.

Two international professionals, Bruno Thurlimann and A. G. Davenport, were added to the team, one an expert on steel structures and the other on wind tunnel testing. After extensive analyses of their test results an unexpected problem emerged as a principal concern. Thurlimann concluded that the structure was well braced in the short direction, but was vulnerable in the long direction due to the weakness of braces and the second-order effect of gravity loads subsequent to lateral deflection. While quite unusual, this problem was also resolved by further strengthening the bracing of the frame in the long direction.

However, all of this still left the glass breakage problem unresolved.

Findings About the Plate Glass Breakage Problem

The root cause of the glass breakage problem was discovered through the method of elimination. Once the foundation, structural frame, and motion problems were eliminated as potential causes of the glass breakage problem, attention was focused on the construction of the cladding system itself.

A group of experts, including engineers William LeMessurier and Victor Mahler, began to study the problem through by monitoring the site conditions, conducting wind tunnel experiments, and studying materials used in the construction of the cladding panel. Their conclusions pointed to the unexpected, the strong bonding that formed between the lead spacer separating the two panes of glass and the reflective silver coating applied to the inner surface of the outer pane. The reflective coating was a part of the effort to neutralize the very tall massing of the tower by turning it into a surface that reflects the contextual buildings around it.

Because of the unintentional bonding of the outer glass pane to the lead spacer, these independent materials started acting as a monolithic, composite material. This resulted in the lead spacer developing fatigue since the flexing of the pane of glass due to wind effects was directly transferred to the lead. While glass could withstand the flexing, the material properties of lead resulted in the development of fatigue. Once this happened, cracks that formed in the lead spacer caused differential bending in the glass.

Eventually the exterior glass panes started showing fatigue as well and cracks appeared in the glass. Subject to continuous and variable wind loads, the cracks telegraphed throughout each vulnerable pane, ultimately resulting in the pieces of glass separating from the rest of the cladding and falling down onto the pavement. This behavior was aggravated during strong wind events such as the one that occurred during January 20, 1973, causing dozens of panes to break in the span of a short period of time (Fig. 12.3).

Resolution of the Glass Breakage Problem

The reason why this problem was not anticipated and not easily discovered was largely due to the fact that this was one of the very few times, if ever, these specific materials were used together, especially the silver coating and the soldered lead spacer in constructing such large panes of glass (Fig. 12.4). Once the problem was identified, the solution was quite clear, but extremely costly: all 10,344 glazed cladding modules had to be replaced by an equivalent assembly without double glazing

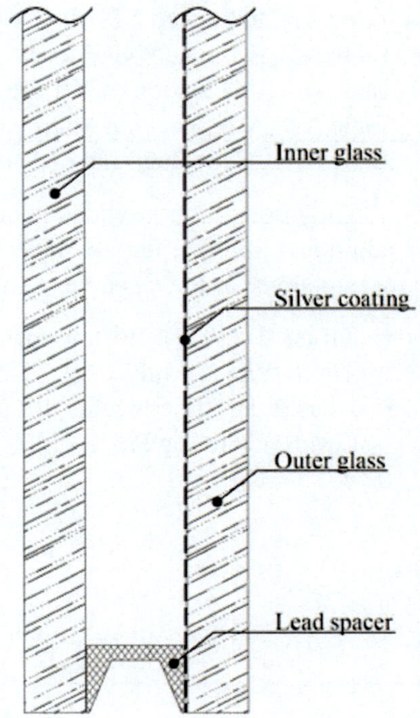

Fig. 12.4 Detail showing double glass, lead spacer, and silver coating. (Courtesy of ADA, Inc.)

and the unusual combination of spacer-materials. In addition to the cost of replacing the cladding, the tower no longer had its energy saving, temperature conserving double-glazing system.

The groups of experts who examined this issue and finally solved the problem were working under a nondisclosure agreement, in perpetuity. The only two individuals who, perhaps by some fortunate oversight, had not signed this agreement were LeMessurier and Mahler. Twenty years later, they came out on their own and explained the mysterious circumstances that surrounded the Hancock Tower glass breakage problem.

Exercise 11

Use one or more of the tables included in the Appendix section to conduct a partial DAV analysis of the case study provided in this chapter. Did the value adding features contribute to cost, schedule, or quality of the product positively or negatively? What could or should have been done to realize the full benefit of the value added by the special features of this design?

Table A.1: Building Facts
Table A.2: Stakeholders
Table A.3: Design Features
Table A.4: Timeline
Table A.5: Estimated ordinal values for Stakeholder Benefits
Table A.6: Estimated NPV for Feature; *Energy Savings*
Table A.7: Estimated NPV for Feature; *Structural System*
Table A.8: Calculation of Net Present Value of a Double-Skin Facade
Table A.9: AHP Ranking Matrix Technique Applied to Similar Buildings
Table A.10: Benchmarking of N-number of Comparable Buildings

The information provided about this case is at best limited. Feel free to collect additional information to undertake a complete analysis and make necessary assumptions. Some resources to help you get started include the following:

- Kifner J (1973) Boston tower's plywood windows. *The New York Times, ARCHIVES*, 18 July 1973
- Aldrich HP, Lambrechts JR (1986) Back Bay Boston, Part II: Groundwater Levels. Civil Eng Pract 1(2).

Chapter 13
Kansas City Hyatt Regency, Kansas City, MO, USA

On July 17, 1981, two suspended walkways within the atrium area of the Hyatt Regency Hotel in Kansas City, MO, collapsed, leaving 113 people dead and 186 injured. In terms of loss of life and injuries, this was the most devastating structural collapse ever to take place in the United States, not including collapses caused by malicious agents like the Oklahoma City Bombing or the 9/11 Attack on the Twin Towers of New York".

On July 20, 1981, the National Bureau of Standards (NBS) was asked to investigate this case and determine the probable cause of the accident. After extensive laboratory tests and onsite inspections NBS issued a report summarizing its findings. They concluded that the walkways that collapsed were designed and constructed in a substandard manner directly contributing to the collapse. No other factor was linked to the disaster (Marshall et al. 1982).

Background

On July 17, 1981, Senator Thomas F. Eagleton's office contacted the NBS and requested that technical assistance be provided to Kansas City. Shortly thereafter Kansas City Mayor, Richard L. Berkley asked NBS to determine the most probable cause of the walkways' collapse. Senators Thomas F. Eagleton, John C. Danforth, and Congressman Richard Bolling endorsed the Mayor's request for the NBS to conduct an independent investigation.

NBS researchers first arrived in Kansas City on July 21, 1981. During the course of the investigation, the NBS team of engineers and scientists inspected the Hyatt Regency atrium area and the warehouse where the walkway debris was stored. In the early phases of the investigation, NBS involvement was limited by court order to visual and photographic observations and measurements. NBS requested permission to weigh selected walkway spans and to remove certain portions of the walkway box beams, hanger rods, concrete decks, and nuts and

Ö. Akın, *Design Added Value*, https://doi.org/10.1007/978-3-030-28860-0_13

Fig. 13.1 Overview of the collapsed walkways at the Kansas City Hyatt Regency (Marhsall et al. 1982)

washers for additional study and destructive examination. The Bureau's requests were eventually granted under court orders after agreement was reached with litigants involved in various legal actions stemming from the collapse of the walkways (Fig. 13.1).

Documents such as drawings, specifications, inspection reports, test reports, and construction logs as well as photographs and videotapes became available to NBS from a number of sources. These were useful in gaining information about the walkways as originally approved for construction and as modified during the construction process.

NBS conducted extensive laboratory tests on Bureau-fabricated mockups of parts of the walkways. Tests were also conducted by NBS on portions of the debris. Analytical models were developed to predict the response of the walkways to various loading conditions. Also, tests were conducted to determine material properties and weld and fracture characteristics.

Configuration

The Hyatt Regency consists of three main sections: a high-rise section, a function block, and a connecting atrium area. As built, three suspended walkways spanned the atrium at the second third and fourth floor levels and connected by the high-rise and function sections. The second floor walkways were suspended from the fourth floor walkway, which was directly above it. In turn, the fourth floor walkway was suspended from the atrium roof framing, by a set of six hanger rods. The third floor walkway was offset from the other two and was independently suspended from the roof framing, by another set of hanger rods.

In the collapse, the second and fourth walkways fell to the atrium floor, with the fourth floor walkway coming to rest on the lower levels. Most of those killed and injured were either on the first floor level of the atrium or on the second floor walkway. The third floor walkway was not involved in the collapse.

As approved by the Kansas City Codes Administration Office, the plans for the walkways called for a single set of hanger rods (attached to the roof framing) that would pass through the fourth floor box beams and then the second floor box beams. The box beams—made up of a pair of 8-inch steel channels with the flanges welded toe to toe—were to rest on hanger-rod washers and nuts below each set of beams. Under this arrangement each box beam would separately transfer its load directly into the hanger rods.

However, during construction, shop drawings were prepared by the steel fabricator, which called for the use of two sets of hanger rods rather than a single set. One of the hanger rods extended from the fourth floor box beams to the roof framing and the other set extended from the second floor box beams to the fourth floor box beams. Hence, all of the second floor walkway loads were first transferred to the fourth floor box beams, where both that load and the fourth floor walkway load were transmitted through the box beam-hanger rod connections to the ceiling. As indicated by their stamps, the contractor, structural engineer, and architect, all, reviewed the shop drawings (Marshall et al. 1982).

Efforts were made to establish as accurately as possible the loads on the walkways at the time of the collapse. Weighing of selected walkway debris along with measurements taken on concrete cores removed from the walkway decks permitted NBS to determine that the actual walkway dead load (self-weigh) was approximately 8% higher than the design dead loads.

Estimates of the loads due to people present on the walkways at the time of the collapse (live loads) were much more difficult to make. Witnesses made a number of conflicting statements regarding the numbers and activities of people on the walkways just prior to the collapse. However, a television crew had videotaped parts of the walkway minutes before the collapse. By studying this videotape, NBS concluded the 63 people represented a credible estimate of combined second and fourth floor walkway occupancy at the time of collapse (Marshall et al. 1982).

Findings

Based on field, laboratory, and analytical investigations, NBS concluded that:

1. Collapse of the walkways occurred under the action of loads that were substantially less than the design loads specified by the Kansas City Building Code.
2. The ultimate capacity of box-beam hanger rod connections can be predicted on the basis of laboratory test results.

3. Under the action of the loads estimated to have been present on the walkways at the time of collapse, all fourth floor box-beam hanger rod connections were candidates for initiation of collapse.
4. Observed distortions of structural components strongly suggest that failure of the walkway system initiated in the box-beam hanger rod connections at location 9UE (east, middle box-beam, fourth floor).
5. As constructed, the box beam-hanger rod connections, the fourth floor to ceiling hanger rods, and the third floor walkway hanger rods did not satisfy the design provisions of the Kansas City Building Code.
6. The change in the hanger rod arrangement from a continuous rod to interrupted rods essentially doubled the load to be transferred by the fourth floor box beam-hanger rod connections.

Exercise 12

Use one or more of the tables included in the Appendix section to conduct a partial DAV analysis of the case study provided in this chapter. Did the value adding features (suspended walkways) contribute to cost, schedule, or quality of the product positively or negatively? What could or should have been done to realize the full benefit of the value added by the special features of this design?

Table A.1: Building facts
Table A.2: Stakeholders
Table A.3: Design features
Table A.4: Timeline
Table A.5: Estimated ordinal values for stakeholder benefits
Table A.6: Estimated NPV for feature; *Energy Savings*
Table A.7: Estimated NPV for feature; *Structural System*
Table A.8: Calculation of net present value of a double-skin Facade
Table A.9: AHP ranking matrix technique applied to similar buildings
Table A.10: Benchmarking of N-number of comparable buildings

The information provided about this case is at best limited. Feel free to collect additional information to undertake a complete analysis and make necessary assumptions. Some resources to help you get started include the following:

- Marshall RD (1982) Investigation of the Kansas City Hyatt Regency walkways collapse. Building Science Series. 143. U.S. Dept. of Commerce, National Bureau of Standards. Retrieved February 17, 2018.
- Petroski H (1985) "Accidents waiting to happen". To engineer is human: the role of failure in structural design. New York: Random House. pp. 85–93.
- Levey M, Salvadori M, Woest K (1994) Why buildings fall down: how structures fail. W. W. Norton & Company.

Chapter 14
Pruitt-Igoe, Saint Louis, MO, USA

The Pruitt-Igoe public housing project in St. Louis, Missouri (MO), occupies a significant niche in the annals of architectural "riches to rags" stories. Designed for low-income families, the complex comprised 33 high-rise, 11-story, buildings containing 2870 apartments on 57.5 acres. As with most "conventional" public housing of this period, that were built, owned, and managed by a government agency, Pruitt-Igoe was part of the larger post-World War II slum clearance and urban renewal efforts launched by the partnership of local municipalities and the federal government (Delatte 2009).

The renewal efforts were aimed at the decayed, poverty-ridden slums of central cities where severe overcrowding and considerable housing stock deterioration existed. The initial effort at Pruitt-Igoe was nothing short of the heroic, and aimed to turn this inner-city slum into a modern suburb. Consistent with this approach, most public housing projects like Pruitt-Igoe tended to be built in a ring within a few miles of the central business district but failed to escape the wrath of the forces of slum formation. In its final irony, the rubble from Pruitt-Igoe was trucked to a giant rock quarry in suburban St. Louis, adjacent to one of the most affluent fringe area communities in America.

Idiosyncrasies of American Public Housing

Several facts concerning conventional public housing in the USA are essential to an understanding of the problems that Pruitt-Igoe faced despite efforts to embellish it with the most advanced design features of the times. Many of these efforts were inspired by the successes of Modern Architecture designed and built for some of the most advanced nations of the globe.

Discounted in this picture is the basic fact that the "free enterprise" capitalism of the American political and social order resisted governmental ownership and management of housing over the past half-century. By 1975, more than 40 years after the first

Ö. Akın, *Design Added Value*, https://doi.org/10.1007/978-3-030-28860-0_14

housing program was inaugurated by the federal government, less than 2% of the nation's dwelling units were publicly owned and operated. Housing provided directly by the government prior to World War II was confined to a relatively few thousand units for defense-related industrial workers and to scattered experimental efforts to supply shelter for families in industrial centers during the economic depression of the 1930s.

Following World War II, America faced a housing shortage of huge proportions. Large-scale migration of rural workers to the industrial cities during the war and the reactivation of family initiation interrupted by military service produced severe overcrowding and heavy demand for housing in the postwar period. Responding to this demand, American private industry built millions of single-family homes in the suburbs.

Low construction costs and interest rates were made possible by the backing of Federal Housing Administration and a burgeoning postwar economy was facilitated in turn by the massive exodus of young families out of central cities into the suburban tract housing. This exodus was followed by a brand new migration to return to nation's core cities. Blacks and whites of lower socioeconomic status from rural farming communities were in search of jobs, better housing, and the "good life" of the metropolis.

It was during this same postwar period that the USA embarked on its first really large-scale building program of public housing through the partnerships between national and local government entities. Unique to the American federal system, this partnership involved a complex arrangement whereby the national government subsidized municipal housing authorities to purchase and clear deteriorated slum areas of central cities, called *Urban Renewal* ever since its inception.

In order to replace at least some of the housing destroyed in this process, the Federal Government provided the funds for local authorities to construct apartments for low-income urbanites. These buildings were to be owned, maintained, and managed by local agencies and financed principally by revenue from rent. This program was not designed as a basic national commitment to provide housing for the poor; the legislative goal, in a concession to the free-enterprise system, remained that of stimulating the economy. This was to be achieved by giving the "*temporary, but hopefully upwardly mobile*" poor a foothold from which to obtain jobs and sufficient economic gains to enter the private housing market.

Public housing was not intended for the poorest of the poor, but for those who had at least some income and could afford a modest payment of rent. Consistent with this general philosophy, maximum income standards were also set, dictating that residents must move out when they achieved income beyond what could be considered a rough equivalent to that of the lower working class. Given this background, it is not too surprising that a project like Pruitt-Igoe could emerge from an architect's drawing board in 1950 as though designed by Le Corbusier for an idealized working class in an Algerian or European city.

The Effort to Salvage

Pruitt-Igoe was the fourth conventional public housing project undertaken by the St. Louis Housing and Urban Development (HUD) Authority. Originally conceived of as two separate projects—one for blacks, the other for whites—the two were operationally combined when the U.S. Supreme Court invalidated the legal principle of "separate but equal facilities" for the races in 1954.

At no time did the project enjoy an occupancy rate of more than 90%, and this level was maintained only for the first several years of its existence. During the 1960s, the occupancy rate dropped precipitously, until by the end of the decade slightly more than one-third of the units remained occupied. Well before this time, the project had fallen into a state of severe physical disrepair and social deterioration, marked by excessive rates of crime and vandalism. Less than 6 years after it was built, it had begun to resemble a disaster area, *"a disgrace, a concrete ghetto, a nineteenth-century caricature of an insane asylum."*[1]

Because of the loss of tenants and the growing problems of security and maintenance, the Housing Authority in 1970 vacated 27 of the buildings and moved the remaining residents into the other six structures. Three years later, it voted to shut down the project entirely, and shortly thereafter, in August of 1973, the mayor announced that the City and HUD had reached agreement to raze all of the buildings. In making the announcement, the mayor said that the demolition would eliminate *"the terrible condition of Pruitt-Igoe and the stigma which its presence has brought upon our city."* Over the next 9 months all tenants were moved from the project, and in early 1976 wrecking crews began the monumental task of razing the huge complex.

The decision to demolish Pruitt-Igoe was the culmination of a series of unsuccessful efforts undertaken over the years to make the giant project operationally viable. These included both physical and social measures designed to eliminate or ameliorate the problems that had plagued the project. Pruitt-Igoe in effect became a target for all the major governmental intervention strategies of the 1960s. It was a specifically designated area for the model cities programs; a focal point for social welfare services; the subject of numerous social science studies; and the recipient of substantial funds for physical rehabilitation. These activities are too numerous to delineate here but several examples will serve to indicate their nature and scope.

The federal department of Health Education and Welfare (HEW) in 1962 committed over $5 million of its funds to Pruitt-Igoe in an attempt to alleviate the social problems associated with the project. This money was to be used over a four-year period for an intensive program of health, education, vocational rehabilitation, and welfare. With the aid of this grant the Missouri state Division of Welfare established an office in the project and assigned a staff of 45 social workers to provide on-site services to the residents. A day care center, health clinic, and other facilities were set up, and a local university conducted a federally funded study of the tenants, worth $750,000.

[1] St. Louis Post-Dispatch, October 29, 1966.

These efforts at social amelioration were accompanied by measures to improve the physical plant. In 1965, HUD allocated $7 million to renovate the project including general repairs, structural changes, installation of a new lighting system, and creation of picnic and playground areas. Seven years later, it authorized another $5 million for physical rehabilitation part of which was to be used to experimentally determine whether high-rise structures could be converted into three- or four-story buildings by eliminating their top floors.

By this time, virtually all concerned—tenant groups as well as public authorities—were convinced that Pruitt-Igoe had either to be radically restructured and reduced to a smaller size or vacated and demolished. Proposals to do the former, including a carefully drawn action plan authorized by HUD and developed by the architectural firm of Skidmore, Owings, and Merrill, evoked little enthusiasm among community leaders. Doubts as to the economic feasibility of revitalizing the project were too strong to be overcome. The seemingly inevitable occurred in 1976 with the razing of the project.

Causes of Failure

Why did Pruitt-Igoe fail? Why did it become, as one newspaper commentator called it, *"the biggest nonmilitary fiasco in the nation's history."* Conventional public housing in the USA has won few accolades; many Americans, in fact, consider the overall program a failure. Yet, in no instance have any of the nation's major projects become the total physical and social disaster that was the fate of the St. Louis complex. No definitive answer to this question has yet emerged although many reasons have been advanced to explain the failure. These generally fall into four categories: (1) physical design; (2) site location; (3) tenant characteristics; and (4) management practices. Depending on who is doing the explaining, one or other of the categories is usually stressed as the prime cause (Delatte 2009).

Physical Design

Severe criticism has been directed at the design of Pruitt-Igoe and at the alleged insensitivity of the architects and planners to the needs and values of the people who were to occupy the units. The policy makers, as well as the designers, must share the blame. The original plans, drafted by architect Minoru Yamasaki, called for a combination of garden apartments and high-rise structures at a density of 30 units per acre. For reasons of economy, this design was significantly altered at the insistence of federal authorities to eliminate the low-rise buildings and increase the density to 50 units per acre.

Equally importantly, substantially higher than anticipated bids led to successive changes in other aspects of the project. These cutbacks included reducing the amount of living space provided for the tenants, dropping virtually all landscaping, eliminating public restrooms at ground level thus encouraging youngsters to use stairwells and breezeways as substitutes, leaving the block walls of the stairways unpainted and exposed uninsulated steam pipes, and skimping on recreational facilities. As Eugene Meehan (1975, pp. 64–65) observed in his study of public housing policy: "*The effort to reduce costs (in Pruitt-Igoe) focused heavily on those dimensions of the project which most directly involved the everyday life of the tenants and which were almost guaranteed to maximize their frustration and discomfort*."

The 33 buildings, which comprised the complex, were identical in design and ordered in rows. The design concept was in the style of Le Corbusier's "high city" reaching out to the sky for light and air. It was little related, however, to the architecture of St. Louis or the living patterns of low-income families. Constructed of reinforced concrete faced with bricks, each building was organized around a single elevator and staircase in the center and stairwells at each end. Automatic skip-stop elevators opened only on every third floor, thus permitting a wide passageway or gallery, 11 feet deep by 85 feet long, to run across the front of the building at each of the floor stops. The architects envisioned these corridors as "vertical neighborhoods" where tenants from the contiguous floors could congregate and small children safely play.

Imaginative as the notion of creating mini-social enclaves within the buildings may have been, the galleries soon became havens for disruptive juveniles and muggers, "gauntlets" through which the tenants had to pass to reach their apartments. The windowless stairwells, which connected the floors to the galleries, created convenient settings for robberies, assaults, and rapes. What had been a design feature intended to enhance the quality of life for the tenants and enhance value of the design had within a few years become a nightmare.

Significantly, the long passageways and other architectural aspects of the project ran counter to the design features for reducing crime outlined by Oscar Newman in his *Defensible Space* (1972). Among the questionable factors were high density, blind staircases, lack of resident control over persons entering the building, neutral open spaces around the individual structures, and circulation patterns generated by the skip-stop elevator system that provided ample opportunities for criminal acts and mischief.

Perhaps the most important casualty of the economy measures applied in the case of Pruitt-Igoe was the paring down of living space within the individual apartments and the increase in the proportion of one- and two-bedroom units to approximately 60% of the total. Less than 8% of the apartments contained more than three bedrooms. This emphasis on the smaller units proved to be a serious mistake, one that compelled the HUD to crowd large families into already cramped units.

Site Location

Located well north of the principal axis of activity, isolated from the main business district and major centers of industrial employment, and devoid of commercial activities except the marginal corner store type of operation, Pruitt-Igoe's physical decline was not anomalous to the neighborhood in which it was situated. At the time the project was constructed, the area, one of the oldest sections of St. Louis, was already badly blighted and in a state of physical decay. Most of the structures dated to the mid- or late nineteenth century, over two-thirds of them lacked adequate plumbing facilities, nearly a third were overcrowded, and many were in a state of disrepair.

Crime rates were high and dope peddling extensive. The area had long been a port of entry for poor migrants. During the late nineteenth and early twentieth centuries white ethnic groups—principally Irish, German, and Italian—were the dominant occupants, low-income black families succeeded them in subsequent decades, and by the time Pruitt-Igoe accepted its first tenants, the adjacent neighborhood was over 75% black.

In 1950, the one-fourth mile area surrounding the Pruitt-Igoe site contained 5300 housing units with an occupancy rate of 99%. By 1970, the number had fallen to slightly less than 4000, with only 75% of the units occupied, a trend that has since continued unabated. Some observers attribute this decline to the negative impact the project had on its environment. Others contend that the neighborhood was already beyond saving and that at most Pruitt-Igoe accelerated, but did not cause its deterioration. Still others argue that the failure of the project was a reflection of the general conditions of the area in which it was located.

Tenant Characteristics

To many St. Louisians the tenants were the major cause of the collapse of Pruitt-Igoe. In their eyes the project was filled with "unworthy" and "irresponsible" poor, the urban "underclass." That all the occupants were black served only to strengthen this impression. HUD statistics, to the contrary, indicate that Pruitt-Igoe residents did not differ greatly from those in the other St. Louis projects in terms of income, rate of employment, and percentage on public welfare. They do, however, show one significant variation. Pruitt-Igoe from its inception exceeded the other projects in the proportion of female-headed families.

In 1956, it contained 30% of Aid to Dependent Children (ADC) mothers compared to an average of 15% in the others, and in 1960 the corresponding figures were 40% and 23%. Because ADC families tend to be large, Pruitt-Igoe's youth population came to be substantially more sizable than that in its local counterparts.

Given the sterile design of the complex, its grossly inadequate recreational facilities, and the general lack of job opportunities for teenagers, this factor may well have contributed to the high degree of vandalism that the project experienced. Whether it was the determining factor for Pruitt-Igoe's failure is extremely doubtful.

Management Practices

Inept management by the local housing authority has also been singled out as a contributing factor to the project's rapid decline. When considering this element, however, the parameters within which the Authority was compelled to operate must also be taken into account. Its maintenance controls, for example, were lax and inefficient and its relationships to the tenants generally poor. Yet, its most damaging policies and practices were either imposed by federal legislation or the result of factors over which it had little control.

One of the most severe limitations on the Authority was its almost total dependence on rent for its income. It was during 1969 that the Housing Act was amended to limit rental to 25% of the tenant's income, with the federal government making up the difference between that amount and the fair market price. With full occupancy and a tenant population having sufficient income to pay economically viable rents, such a policy may have worked. In the case of Pruitt-Igoe, neither of these conditions was present. At no time after the first few years of its existence did the project produce more than 40% of the rentals needed for successful operation (Meehan 1975). Moreover, high maintenance costs, aggravated by extensive vandalism, drained funds from the financially healthier projects to the eventual detriment of all. In efforts to remain solvent, the Authority was compelled to raise rents, cut maintenance, and become less selective in screening new tenants.

Certainly, the Authority can hardly be charged with "writing-off" the project until all hope had dissipated. The numerous efforts at rehabilitation run counter to such a contention. As the executive director of the agency observed in drawing a parallel with American policy in Vietnam: "*Always there was something we thought we could do to salvage the disaster. It was ingrained; we kept going back in with more money.*"[2] Given the surrounding circumstances and the constraints within which the Authority functioned, it is questionable whether the eventual outcome would have been materially altered by any changes in local administrative practices or policies.

[2] National Observer, April 3, 1970, p. 7.

Exercise 13

Use one or more of the tables included in the Appendix section to conduct a full or partial DAV analysis of the case study provided in this chapter. Did the value adding features contribute to cost, schedule, or quality of product positively or negatively? What could or should have been done to realize the full benefit of the value added by the special features of this design?

Tab.le A.1: Building facts
Table A.2: Stakeholders
Table A.3: Design features
Table A.4: Timeline
Table A.5: Estimated ordinal values for stakeholder benefits
Table A.6: Estimated NPV for feature; *Energy Savings*
Table A.7: Estimated NPV for feature; *Structural System*
Table A.8: Calculation of net present value of a double-skin Facade
Table A.9: AHP ranking matrix technique applied to similar buildings
Table A.10: Benchmarking of N-number of comparable buildings

The information provided about this case is at best limited. Feel free to collect additional relevant information to undertake a complete analysis and make necessary assumptions. Some resources to help you get started include the following:

- Hoffman A (permanent dead link) "Why They Built the Pruitt–Igoe Project" Joint Center for Housing Studies, Harvard University.
- Montgomery R (1985) "Pruitt–Igoe: Policy Failure or Societal Symptom" *The Metropolitan* Midwest Urbana. Chicago: University of Illinois Press.
- The Pruitt-Igoe Myth (2011) The Documentary Movie.

Chapter 15
Crystal Palace, London, UK

Sir Joseph Paxton (1803–1865), a self-made botanist, designer, and engineer of greenhouses, and a recipient of Knighthood due to his accomplishments, was the principal designer of the Crystal Palace, the structure inside of which the Great Exhibition of 1851, a world's fair of technology and production of the times, was held. At the time, this was the largest international exhibition held anywhere. It marked the culmination of an extraordinary confluence of forces emanating from international commerce, technology, architecture, and politics. Its impact on the engineering of buildings, not to mention the architectural world is, surprisingly, remarkable (Hobhouse 1950).

By some miracle of history, the first great exhibition glorifying modern industrialism was housed in a structure, which more perfectly expressed its potentials than any that ever followed it (Fig. 15.1).

Background

The Great Exhibition of 1851, whereas it turned out to be a spectacular success, did not start out with that promise. Having been upgraded, primarily for political reasons, to the level of an ambitious international event from that of a relatively small-scale national exhibition, it suffered from a severe lack of advanced planning. This was particularly true in the case of the architectural design of the building to house it.

This design was obtained at the last minute, at a time when all options were practically exhausted. While a design competition, which yielded more than 230 entries, was held in advance, the jury decided to use none of these designs, including that of the winner.

As the jury, in a desperate move, was contemplating to assemble a hybrid design from the outstanding features of some of the competition entries, Mr. Paxton was given a chance to submit a design for the building. Mr. Paxton found himself at the right time, in the right place, and with the appropriate knowledge and skills to

Ö. Akın, *Design Added Value*, https://doi.org/10.1007/978-3-030-28860-0_15

Fig. 15.1 Exterior view of the Great World' Exhibition building of 1851, the Crystal Palace (Image credit: "Gran incendio del Crystal Palace (Londres 1936)" by Recuerdos de Pandora is licensed under CC BY-SA 2.0)

handle this problem; it can be said that, serendipitously, his entire life's work became relevant to the problem at hand.

Paxton started in the field of horticulture at the age of fifteen. Later, in 1823, he was employed by the London Horticultural Society to work on the Cheswick gardens of the Duke of Devonshire. In this capacity, he was exposed to both the field techniques and the organization of botanical and horticultural knowledge. Eventually, he rose to the post of the superintendent of the Duke's countryseat at Chatsworth. Throughout his tenure there he nurtured his interest in the understanding of plants and organic life and built numerous structures in which to houseplants. He also traveled abroad and served as the editor of horticultural journals.

When he was given the chance to design the structure for the Great Exhibition, he was probably one of the most knowledgeable men anywhere about the construction of light structures and the organization and supervision of their construction. In his earlier designs, he had perfected a wooden glazing system, called the "ridge-and-furrow" system, and a variety of roof forms and spans with glazing materials. In these designs he appears to have developed a structural understanding of natural forms. For example, the leaves of the Victoria Regia served him well in integrating many design requirements in a singular formal solution.

The design for the World's Fair was completed in eight days. It responded to several important design criteria through several design features: lightness, de-mountability and re-mountability, large spans, flexibility of use, and, most severe of all, completion of the entire construction within 26 weeks or six months.

Based on the design he provided, we might add that the building was also modular, prefabricated, with integrated systems of condensation drainage, structure and a construction assembly system. It was the largest enclosure to date, boasting 989,884 sq.ft. of space and one and a half miles long balcony space. Owing to the prefabrication technique as well as special assembly methods he developed the construction was completed on time.

Achievements of the Crystal Palace

Both in terms of design and construction the accomplishments of the building are quite remarkable. In addition to having the largest interior space of the day (Fig. 15.2), it provided the most advanced vision for a structure designed to house the future technologies of the time. Not only was it modular for ease of relocation, but it also introduced standard manufactured details, complete prefabrication, systems integration, and site assembly procedures with industrial levels of efficiency, all of which were concepts beyond their time.

The building was modularly designed and fabricated. Its assembly at the site was part of the original design consideration. Parts were manufactured and shipped to the site in the order of assembly, preconceiving the idea of fast-tract construction, which would emerge a full century later. New methods for assembly at the site had to be invented. Some prefabricated members were too large and heavy to be assembled in the conventional ways. Rigs, which can only be identified as the predecessors of the modern construction crane, was developed (Fig. 15.3).

A rail-cart was fashioned by Paxton based on his ridge-and-furrow design developed decades ago to make glazing assembly more efficient. This was just large enough to carry a set of panes of glass and four workers. It fit exactly on the framing of the glazing system and used the framing as its rails. These channels that served as rails during construction served as gutters that sent rainwater and interior condensation to down spouts that were strategically placed throughout the structure, once the construction was complete.

Not only in terms of technology, but also in terms of spatial composition Paxton's design predates some of the innovations that emerged only in the mid-twentieth century. Being a botanist by training, Paxton was motivated to save the natural flora of the site. This included large trees some of which in the end became indoor plants adorning lofty interior enclosures, reminiscent of the modern atria of our day. Mezzanines surrounded some of these spaces adding to their vertical sense.

The most remarkable lesson that one can draw from the Crystal Palace is not the fact that these advances were made but why were they made in the first place? There is no easy answer to this question. Scholars of the building converge around several points.

Fig. 15.2 The interior of the Crystal Palace

Due to his technical background, Paxton approached the problem as an architectonic-engineering problem, creating novel forms borrowing from the annals of building technology. He was able to adapt a technology that he knew intimately and in which he personally had a hand in as an innovator for decades. He happened to live at a time when new building materials, styles, and techniques were emerging. He was bold enough to push these avant-garde ideas to the forefront in this work (Fig. 15.4).

Ultimately, Paxton understood the challenges and limitations of the new problem and focused on solving these directly and almost single-mindedly. He understood that if he were to solve the problems of timetable and enclosure in this building, he had to innovate.

He defined these problems so that they could be solved using all means available at the time, regardless of however unorthodox they may have seemed at the time. There was very little room on his design schedule to refocus the problem on a new issue. All he had time for was a single shot at perfecting a unique concept, for a unique problem.

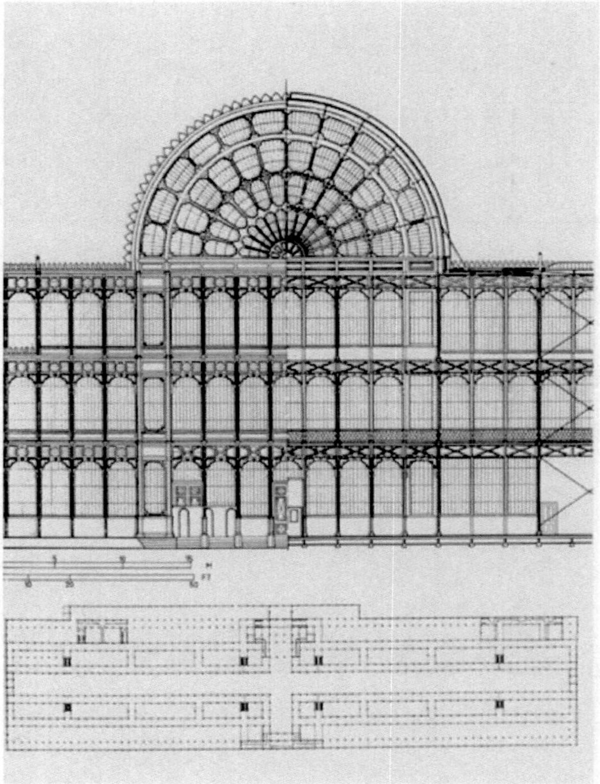

Fig. 15.3 The modular design of the Crystal Palace

The End

In hindsight, the one thing that could have been, but was not a part of the design objectives was fireproofing. The building was not fireproofed in the contemporary sense of that word, since at the time it was believed that a structure primarily made from iron and glass would be substantially fire proof. This turned out to be a fatal omission. After the conclusion of the world's fair the structure was demounted and remounted at a different location. Adapting its footprint and form to the new context, the buildings innovative design was demonstrated once again.

Almost 85 years to the day after its construction, at its new site, the structure burned down in a spectacular and brief fire, in 1936. Consequently, we can study it only through the records left from its days of glory.

Fig. 15.4 The atria-like interior of the Crystal Palace (Image credit: "Steel Engraving; Crystal Palace, 1851 Exhibition Welcome" is licensed under CC BY 4.0)

Exercise 14

Use one or more of the tables included in the Appendix section to conduct a partial DAV analysis of the case study provided in this chapter. Did the value adding features contribute to cost, schedule, or quality of the product positively or negatively? What could or should have been done to realize the full benefit of the value added by the special features of this design?

Table A.1: Building facts
Table A.2: Stakeholders
Table A.3: Design features
Table A.4: Timeline
Table A.5: Estimated ordinal values for stakeholder benefits
Table A.6: Estimated NPV for feature; *Energy Savings*
Table A.7: Estimated NPV for feature; *Structural System*
Table A.8: Calculation of net present value of a double-skin Facade
Table A.9: AHP ranking matrix technique applied to similar buildings
Table A.10: Benchmarking of N-number of comparable buildings

The information provided about this case is at best limited. Feel free to collect additional information to undertake a complete analysis and make necessary assumptions. Some resources to help you get started include the following:

- Hobhouse C (1950) 1851 and the Crystal Palace. John Murray, Albemarle Street: London.
- McKean J (1994) Crystal Palace: Joseph Paxton & Charles Fox. Phaidon Press: London.
- Chadwick GF (1961) Works of Sir Joseph Paxton. Architectural Press.

Chapter 16
Sydney Opera House, Sydney, Australia

One of the most celebrated modern buildings in the world, next to the Fallingwater, must be the Sydney Opera House. From its inception to its implementation, it has occupied the attention of the Australian nation as well as that of the Western world. It has been one of the most complex and visible construction projects anywhere in the world. The design created by the Danish architect, Jorn Utzon, has had no equals, or successor for that matter. Its implementation was possible only because an unprecedented national and professional resolve overcame sizable political, technical, aesthetic, financial, and cultural obstacles (Messent 1997).

A Brief History

During the postwar years, the Sydney Opera House became the symbol of Australian desire to create a world class cultural institution for the performing arts and in the process elevate the artistic consciousness of the Australian common folk to higher levels. This idea predates the postwar years and can be observed in the actions of the first General Manager of the Australian Broadcasting Company, Mr. Charles Moses, and the first world renowned conductor of the Sydney Symphony Orchestra, Eugene Goossens, who was appointed to his post in 1946.

There was no doubt that all major branches of the performing arts in Australia, opera, ballet, symphony, and theater, desperately needed a shot in the arm. These activities were principally held in make shift, retrofitted buildings with substandard staging facilities and small audience halls. Thus, the principal motivation for new buildings included increasing capacity and creating first-rate back stage facilities.

By the time early fifties rolled along, there was public awareness of the problem and politicians had latched on to the idea as one that carried cultural and political weight. In fact, Premier of New South Wales Joe Cahill, in 1954, became one of the principal supporters of the project to the extent that the rising and falling fortunes of

Ö. Akın, *Design Added Value*, https://doi.org/10.1007/978-3-030-28860-0_16

the building became a barometer for the faith of his government for a number of years to come.

The role that this building and its design played in the life of an entire nation for a period of a decade is absolutely unprecedented in modern architectural history. After extensive public and private debate, on November 30, 1954, the Australian government under the leadership of the Premier Cahill decided to establish an opera house in Sydney: *"As you are aware, the Government has decided that an Opera House shall be established in Sydney and that it will be worthy of this city... The people who have gathered here this morning are interested in opera and other forms of art to which perhaps many in the community might not give a second thought. Expertise has proven that it is always the few who have to pioneer the way, and the Government has decided to go ahead with this project."*[1]

What came next is as much the evolution of a brilliant design as it is that of a nation's artistic consciousness, all played out in the public and private forums of Australian political institutions.

The Competition

A group of administrators, architects, planners, and musicians decided that the design should be obtained through an open international design competition and the brief of the building should contain the following activities:

"1. Symphony concerts; 2. Choral Festivals; 3. Operatic Performances; 4. Ballet; 5. Drama; 6. Small orchestral concerts, i.e., chamber music, etc..."[2]

More specifically, the program for the building consisted of the following requirements:

"There shall be two halls—one large and one small hall.

A. *The large hall should seat between 3000 and 3500 persons. The large hall to be designed for use for the following purposes:*

    1. *Symphony Concerts.*
    2. *Large-scale Opera.*
    3. *Ballet and Dance.*
    4. *Choral.*
    5. *Pageants and Mass Meetings.*

B. *The small hall should seat approximately 1200 persons. The small hall to be designed for use for the following purposes:*

    1. *Dramatic Presentations.*

[1] Messent, David (1997) Opera House Act One David Messent Photography, Unit 17 28 Rosenberry Street, Balgowlab, NSW 2093, Sydney Australia, pp. 73.

[2] Ashworth papers in Messent (1997), pp. 87.

2. *Intimate Opera.*
3. *Chamber Music.*
4. *Concerts and Recitals.*
5. *Lectures*".

The requirements under items 1 and 2 above have been listed in order of priority with respect to the attention, which should be given to their specialized building needs.

It is expected that ideal conditions will be provided as far as possible acoustically, visually, and in connection with stage and orchestral facilities. Compromises which will prejudice the entirely satisfactory performance of a function with a higher priority in the above list should not be made."[3]

In addition, the competition document included information about (1) the site that was located at Bennelog point in Sydney harbor, (2) local climate, and (3) the prize money of 1st prize = £5000; 2nd prize = £2000; and 3rd.prize = £1000. No budget was provided, and no promises were made regarding the unaltered implementation of the winning project.

Submissions

At the time of the opening of the international competition, 721 requests of the competition documents were made. Seventy-nine of the potential competitors asked 470 questions about the terms of the competition. In December 1996, 270 submissions were at hand to be reviewed by the four-man international panel of architects: Leslie Martin, Cobden Parkes, Eero Saarinen, and Harry Ashworth.

Entries came from 27 countries including Australia, the UK, the USA, South Africa, Denmark, Czechoslovakia, French Morocco, Israel, and Cyprus. Saarinen arrived 2 days late, on January 11, 1957, and joined the proceedings in progress. In reviewing the projects discarded by the other members of the panel he came across the project submitted by Utzon (Fig. 16.1).

Utzon's project violated the competition requirements by going beyond the site boundaries, particularly on the West side of the site. The elevations of the building were imprecise and included poorly detailed photo enlargements of smaller free-hand drawings.

These drawings were clearly submitted just to meet the minimum drawing requirements and were necessitated by lack of time. Yet, Saarinen made an impassioned plea to the rest of the jury and Utzon's project was re-admitted into the short list of projects being considered. Saarinen continued to push this scheme and became its ardent defender against the queries of the other panelists. He even offered

[3] Denis Wolanski library in Messent (1997), pp. 91–92.

Fig. 16.1 The Sydney Opera House at Sydney Harbor (Image credit: "Sydney opera house" by jimmyharris is licensed under CC BY 2.0)

perspective drawings of his own to supplement the submission. His persistence paid off. In the end, Utzon's project was given the first prize.

It was explained later that Saarinen was moved by the vision of this scheme due to its sensitivity towards the site and the harbor area within which it was located. Furthermore, it is conjectured that Saarinen, having just completed a shell-structured amphitheater at the MIT campus in Boston, was drawn to the shell structure of the scheme proposed by Utzon.

It was clear that the project would have to be further detailed before it could be constructed. There was concern, even by one of the jurors, that the construction of the shell structure as proposed was infeasible. Early estimates indicated that the project would cost £4,800,000 and would take 5 years to complete.[4]

In spite of practical problems, the design and its architect achieved instant and worldwide notoriety. Many hailed the project as visionary while others ridiculed it due to its iconographic appearance. Funding was most often used as a justification to abandon the project. In the 1958 elections, the key issue for the electorate was indeed the state of the Opera House. Cahill was reelected with a small margin, running on the platform that he would continue to work towards building the Opera House (Messent 1997).

[4] SMH in Messent (1997) pp.170.

During the next few years the construction documents of the project were developed, as fundraising through state run lotteries were allocated to the project. A fortunate development at the beginning of this period was the involvement of Ove Arup and Associates as project engineers. Their experience with shell structures and large international projects turned out to be a key for the eventual success of the Sydney Opera House.

Primarily due to the involvement of the Arup Associates the project was divided into two stages, the podium and the shell structure. Later, new stages were also added: the tiles, interior finishes, and furnishings. The project was completed nearly 12 years and £60 million later, more than a 1000% increase over the initial estimates.

The Podium

The podium upon which the sail-like shells of the building was to be placed constituted the first stage of the construction. This partitioning of the building into two major stages was more than a logical one to organize the construction and to gain time for the design of the shell structure. It was part of Utzon's original conception of the scheme. He states that his inspiration came from the Mayan temples of the Yucatan in Mexico:

> *The platform as an architectural element is a fascinating feature. I first fell in love with it in Mexico on a study trip in 1949 where I found many variations... A great strength radiates from them... By introducing the platform with a level at the same height as the jungle top, these people had suddenly obtained a new dimension in life.. On these high situated platforms—many of them as long as 100 meters—they built their temples.*[5]

Utzon was not building a temple—at least, not literally—but this metaphor served him well in providing a massive and stable foundation from which his dynamic sail-like forms could spring. The visual as well as the functional relationship between the podium and the roof forms is convincing and well conceived, to say the least. The podium contained all of the functional spaces, like traffic and pedestrian circulation, the auditoria, the library, and theater spaces. It also supported the roof structures and auditoria volumes over the major assembly spaces.

As Utzon and Arup were busy refining the design of the proposed sail like, the client was refining his needs for the building. In June 1958, the committee of the building made changes to the administration area, workshops, and dining rooms. On July 2, 1958, it was announced that the Sydney Opera House would now contain an "experimental theater" with 400 seats. Two days later a television broadcasting studio facility was added to the wish list. In September of the same year, the orchestra pit was enlarged to accommodate 120 musicians instead of 105.[6]

[5] Utzon papers in Messent (1997), article for Zodiac, pp. 157-15.

[6] Ashworth in SMH and Arup papers in Messent (1997), article for Zodiac, pp. 160.

Construction officially began in August 1958 by the demolition of the existing structure at Bennelog Point. The contractor for Stage I, Civil and Civic was the lowest bidder with £1,397,929, 40% less than the highest bidder. Some thought that this was a deliberate underbidding by Civil and Civic in order to gain the contract for this very visible project. The completion time of 18 months was more than doubled by the time Stage I was substantially complete.

There were several reasons for the delay including unexpected foundation conditions, incomplete construction documents and design revisions of the podium as well as the upper structure, which continued almost all the way till the very end.

In March 1961, with the podium only three quarters complete, contractors Civil and Civic complained in writing that this "*contract has been a long history of changes, frustrations, inability to logically plan the work due to the lack of information and detail.*." Indeed, after 21 months into the job, only 55 percent of the construction drawings were completed. These drawings constituted a total of 700 drawings with 695 amendments. Of the 1500 items in the original provisional bill of quantities, 500 were never used and a new 500 new ones were generated.[7]

One of the critical areas of design, redesign, testing, and deployment was the concourse area which was spanned by 50 pre-stressed folded beams, each one 6′ wide and up to 4′ 6″ deep. With some spans exceeding 160′, the supports at midpoint for these beams—whose cross sections changed continuously from a T-shape to a rectangular-shape—shown in the original design were subsequently removed.

The foundations consisted of 460 piers of varying depths (20′–70′) depending on the variety of soil conditions some of which were discovered only during construction. The design called for occupiable spaces below the water level (the orchestra pit being 15′ below) leading to excavation and water proofing problem during construction and occupancy.

There were several important reasons for the litany of design changes. First, Utzon worked with a very high set of design standards and held the belief that design never ends, even to the bitter end of falling behind the schedule and the budget. The Australian government did not want to squander the public support behind the project by waiting for years for the perfect set of construction documents. Hence the project was fast tracked. The contractor had bid low in order to share the limelight and as a consequence wanted to exploit the delays caused by design revisions to a maximum extent.

In September 1962, with the podium nearly complete the contractor sued for additional payment, to the tune of £3,005,000 total—a 120% increase over the original bid price. In the end, the contractors settled for £2,580,000, roughly a £1 million increase over the original budget of £1,511,000, including already approved increases over the bid price. By this time, the design of the shell structure had gone through one major overhaul and the Stage II was about to commence.

---

[7] Arup papers in Messent (1997) pp.171.

The Shell

The original design of the roof forms was elliptical in geometry and were to be constructed as thin ferro-cement shells. The design of the roof structure continued until 1961, with intense effort to accurately predict the stresses on the structure.

Up to this point the only issue in question was how to build the geometry given in the original design, with minor modifications. The geometry of the shell form was never in question. The design being entirely novel, however, this amounted to a monumental task. There were no analytical methods to design this form. Instead testing on elaborate physical models was required to develop the technology for the structural design of the shell structures. Computer technology was in its infancy at the time, and to complicate matters further, several alternative construction technologies were being considered such as thin shell, ribbed shell, and steel.

Early in 1961, the prevailing opinion on the part of almost all involved began to shift to a spherical geometry and to a ribbed structure. Even Utzon was converted to the spherical geometry. As a matter of fact, he proposed the exact geometric shape for the new "shell" design with his now famous wooden model extracting segments from the same sphere to create the curves of the entire shell language of the building. This was truly a watershed moment for the entire project. It made the design of the upper structure possible while causing a great deal of public uncertainty and at least one resignation of a key staff member from Arup Associates.

The selection of the contractor for Stage II was done differently. The firm invited to carry out the job, Hornibooks Inc., not only had expertise in the field but was also nearly as sophisticated in engineering design as the Arup Associates. During the course of the design phase, Hornibooks Inc., developed several novel computer applications to assist in the calculations. They were by and large responsible for the successful completion of the project, even though the project went through further political, technical, and financial convulsions.

The added dead loads due to the shift to a ribbed concrete shell meant that the support foundations had to be redesigned, demolished, and reconstructed [70′ long with bases 7′-15′ in diameter, with 28′ below ground and 42′ above ground]. The ribbed roof shell had to be broken down into similarly shaped pre-stressed segments to facilitate prefabrication. Since the curvature was spherical this was now possible; in fact, the possibility of prefabrication was the principal motivation to shift to a spherical form.

The shell modules had to be designed to hold temporary pre-stressing during the construction stage since with the curvilinear form of the roof scaffolding was virtually impossible to design. To create lateral bracing between the ribbed modules sophisticated bolting and sealing details were designed. The roof tiles, more than 2 million of them, were also assembled in a modular fashion in order to enable factory production. This complicated the ribbed shell design even further.

Epilogue

To further add to these complications, Utzon resigned his position in February 1966, due to political reasons, prior to the completion of Stage II and commencement of the interiors finishes of the building. The project was completed under the supervision and interior design work of the Australian firm Hall, Todd and Livermore.

Today the building stands as the pride-inspiring symbol of Sydney, and indeed of entire Australia, as it was originally intended. It has a full schedule of artistic performances including theater. The main hall is the exclusive territory of the Sydney Symphony.

This has displaced the opera and ballet functions to the smaller secondary hall. In turn, the theater function has been moved out of both upper halls into rooms carved out of the podium space—a process which started during the construction of the building and that still continues. This led to three, major programmatic complications.

1. Owing to the unique forms used in the design of the building, the seating capacity in the two main halls, symphony and opera, are limited to 2679 and 1547, respectively. This grossly underrepresented the desired capacity of these halls. Furthermore, due to the limitations imposed by the shell form, space in the back stage of the building is severely restricted. Additions and modifications are nearly impossible due to the high stress concrete used as the construction material.
2. The theater function is now exclusively in spaces carved out of the volume of the podium. The capacity and functionality of these spaces are also limited by the high stress concrete construction used in the podium area. As a result, in place of a few large theatrical performance halls there are three with smaller seating capacity: 350, 389, and 530, respectively. Furthermore, the size of the back stage area of these spaces is limited and cannot be expanded further due to the reinforced concrete construction.
3. In order to realize the performance functions of the building—even in the limited way that they are being realized—some other functions of the building had to be sacrificed. Space for office and storage areas is curtailed. Lobby areas are cramped. The library of the Sydney Opera House has been disbanded to make room for one of the most recent theater facilities, donating the collection of the Sydney Opera House to other local libraries.

At the beginning of the entire effort, the Sydney Opera House was justified on the basis that the existing facilities were deficient exactly in these regards: back stage flexibility audience capacity and other support functions. After 14 years and 60 million pounds of expenditure, some of these deficiencies still persist.

On the other hand, Sydney became the proud owner of a world-renowned architectural symbol. Its symphony, ballet, and opera performances are coveted by performers and audience alike. Also, while the architect who gave Sydney its architectural and civic identity left in personal disappointment before the completion of the building; his firm under the direction of his son Jan Utzon has been given a new commission to finish the task started decades ago by his father.

Exercise 15

Use one or more of the tables included in the Appendix section (captions listed below) to conduct a partial DAV analysis of the case study provided in this chapter. Did the value adding features contribute to cost, schedule, or quality of the product positively or negatively? What could or should have been done to realize the full benefit of the value added by the special features of this design?

Table A.1: Building facts
Table A.2: Stakeholders
Table A.3: Design features
Table A.4: Timeline
Table A.5: Estimated ordinal values for stakeholder benefits
Table A.6: Estimated NPV for feature; *Energy Savings*
Table A.7: Estimated NPV for feature; *Structural System*
Table A.8: Calculation of net present value of a double-skin Facade
Table A.9: AHP ranking matrix technique applied to similar buildings
Table A.10: Benchmarking of N-number of comparable buildings

The information provided about this case is at best limited. Feel free to collect additional information to undertake a complete analysis and make necessary assumptions. Some resources to help you get started include the following:

- Messent D (1997) Opera house: act one. David Messent Photography, Sydney, Australia.
- Stübe K, Utzon J (2009) Sydney opera house: a tribute to Jørn Utzon. Reveal Books.
- Watson A ed. (2013) Building a masterpiece – The Sydney opera house – lessons in space and environment. Sydney: Powerhouse Publishing.
- Woolley K (2010) Reviewing the performance the design of the Sydney opera house. The Watermark Press.

Chapter 17
Citicorp Tower, New York City, NY, USA

Citicorp Center building is a 59-story, mid-Manhattan tower well known for its unique urban form. The building is supported by four, nine-story columns located on the center points of the lot lines, rather than the corners that define the block that it occupies (Fig. 17.1). This peculiarity is a result of the deal that the owners had to strike in order to purchase the air rights of the church, which still occupies one of the corners of the site (Fig. 17.2).

It was this unique form that caused a series of events, which eventually lead the principal design engineer of the building to discover a major structural flaw. The discovery was surrounded by a certain drama highlighting important professional and DAV issues. Arguably it was a matter of time, in fact possibly a matter of weeks, before a major event could cause the collapse of the entire building.

The only person who discovered this flaw was LeMessurier, the engineer, and he was under no legal obligation to reveal the problem, particularly not before, as his lawyers would later argue, he covered his liability. In his laudable effort as a creative designer that would have added considerable value to this tower, several professional oversights occurred almost moving the entire value of the tower to the deficit column, along with a number of adjacent buildings in Manhattan. Nevertheless, Mr. LeMessurier went ahead with his disclosure which was followed by a series of remarkable events that helped solve the structural problem and averting the danger to the building as well as the surrounding properties.

Background

The Citicorp building is located on the east side of Lexington Avenue between 53rd and 54th streets. A newly constructed St. Peter's Church, replacing the old one, occupies one of the corners of the block (Fig. 17.2). The structural bracing that rises from the top of the four pillars that hold the building up span the entire length of the 59-story building, for 914′.

Ö. Akın, *Design Added Value*, https://doi.org/10.1007/978-3-030-28860-0_17

Fig. 17.1 The Citicorp Tower (Image credit: "NYFeb07 061" by p_c_w is licensed under CC BY-ND 2.0)

Fig. 17.2 St. Peter's Church to the left, located at the corner of the Citicorp Plaza (Image credit: "Base of the Citigroup Center" by Tdorante10 is licensed under CC BY-SA 4.0)

Architect Hugh Stubbins designed the building. Its design was noted from the beginning due to its very tall support pillars, unique top, and some of the engineering innovations that it possesses. Aside from the unusual location of these support pillars, the cross bracing concealed behind its slick, exterior cladding, and its super-light steel structure; it is also credited with one of the very early, successful applications of a *tuned mass damper.*

The building is entirely of steel frame. Six tiers of eight braces, forty-eight in all, secure the structure against wind loads. The frame of the building and the lateral bracing constitute 25,000 tons of steel compared to 60,000 superstructure of the Empire State Building, which was a principal reason for keeping its cost to $175 million. In 1977, when it was completed, it was the 4th largest structure in the world. To reduce the excessive sway of the tower due to lateral wind forces, a tuned mass damper consisting of two 400–410 ton, free-floating concrete blocks was installed at the canted top floor of the tower.

The Structure

In June 1978, soon after the completion of the Citicorp Building, a student from New Jersey called Mr. LeMessurier inquiring about the unusual framing of the building and its structural safety in light of the location of the four supporting pillars. LeMessurier was confident of his design and explained to the student how the cross bracing helped the building's lateral stability. This innocent exchange led to his decision to use this case as an example in a course he was teaching at the time. Subsequently, he re-calculated the strength of the structure against lateral-quartering winds that particularly apply to buildings with similar support systems and geometry.

He was surprised to find that the quartering winds increased the design loads by about %40 for half of the bracing elements. However, with the tuned mass damper and the strong welded[1] connections between the bracing elements and the building frame, the building was in no danger of failure.

Soon after, LeMessurier was confronted with a similar design question regarding another building for which he served a consultant, in Pittsburgh. He was trying to persuade a potential manufacturer of the building's steel structure, U.S. Steel, to consider a similar bracing system. When he called his office to ascertain the connection details of the Citicorp Building, to his surprise, he found that the welded connections were scrapped in lieu of bolted connections. This was done by his branch office in New York, upon the request of Bethlehem Steel, the steel manufacturer of the Citicorp Tower.

At the time this appeared to be a reasonable decision even though his branch office had failed to inform him about the substitution. The two concerns put together, the design for quartering winds and the substitution of the bolts for the welded

[1] This was LeMessurier's recollection at this time.

connections, raised new concerns in Le Messurier's mind. About a month later, he decided to double-check the design calculations of the Citicorp Building on his own.

To his horror, he found out that not only were the quartering winds not taken into account, the calculations were based on considering the cross bracing as a truss rather than as columns, which according to American Institute of Steel Construction standards exempted them from additional safety factors against high wind loads. Consequently, the bolts installed at the connections were inadequate to withstand the tensile loads that would be produced in a 16-year storm.

LeMessurier faced a dilemma: to disclose or not to disclose. If he did, his reputation and professional life as an engineer would be at risk. If he did not, he was jeopardizing many lives and very large sums of property value. His lawyers were advising him to be cautious, while his wife plus his instincts were pushing him in the opposite direction.

Resolution

Against the legal advice he received and emboldened by Mr. Stubbins support of his decision, LeMessurier revealed the facts to the representatives of the Citicorp organization and their technical and legal consultants. He had a hard time convincing them that there was something wrong. Once this was accomplished, he had an equally hard time convincing them that there was nothing else wrong.

Once this too was accomplished, strategies had to be developed to (1) remedy the problem before the hurricane season hit, (2) to prevent the public and the occupants becoming panicked about the state of the building's structure, (3) figure out a way of fixing the bracing problem, and (4) find a way of paying for it all. The solution to these problems was almost as fantastic as the circumstance that brought it to light.

The harsh weather season was indeed around the corner. In fact, during the repair period one hurricane heading towards New York turned around and went a different way sparing the building and the crews who were feverishly working on the repair job. The repair job was concealed by working in the evenings after the building was empty and by building wooden enclosures around the areas being repaired.

To prepare for the worst-case scenario, local officials and the Red Cross were involved in census counts and evacuation plans to be put into effect in the event of an emergency. All parties were sensitive to the issue of taking all precautions but, in order to prevent panic, not letting the public know about the gravity of the situation. Le Messurier's luck kicked in once again. Just as the press wised up to what was going on all New York papers went to strike for a duration that included the entire repair period.

The actual repair involved the welding of an H-shaped Band-Aid at the joint of the giant cross braces where they met the building frame. This required the removal of all interior finishes and then their reinstallation before the repair task would move up to a higher floor. Also, the tuned mass damper, which would actually help the structural integrity of the building during a storm, was equipped with an emergency generator in order to insure its continuous operation.

When all was said and done, the estimated cost of the entire operation was around $4 million. By some accounts the cost could have been estimated higher or could have been easily inflated by the contractors, since much of it was done under emergency conditions and not all expenditures were carefully added to Le Messurier's debit column. Even with this optimistic bottom line, his insurance covered only half of this sum. In the end, Citicorp agreed to cover the rest.

Le Messurier's forthrightness, diligence, and professionalism had won many over to his side and he was spared the brunt of the damages. In fact, even his insurance company decided to reduce his premiums due to the fact that his actions in fact averted a much larger potential claim for them.

Exercise 16

Use one or more of the tables included in the Appendix section (captions listed below) to conduct a partial DAV analysis of the case study provided in this chapter. Did the value adding features contribute to cost, schedule, or quality of the product positively or negatively? What could or should have been done to realize the full benefit of the value added by the special features of this design?

Table A.1: Building facts
Table A.2: Stakeholders
Table A.3: Design features
Table A.4: Timeline
Table A.5: Estimated ordinal values for stakeholder benefits
Table A.6: Estimated NPV for feature; *Energy Savings*
Table A.7: Estimated NPV for feature; *Structural System*
Table A.8: Calculation of net present value of a double-skin Facade
Table A.9: AHP ranking matrix technique applied to similar buildings
Table A.10: Benchmarking of N-number of comparable buildings

The information provided about this case is at best limited. Feel free to collect additional information to undertake a complete analysis and make necessary assumptions. Some resources to help you get started include the following:

- Brady S (2015) Citicorp center tower: how failure was averted. Engineer's Journal.
- Werner J (2014) The design flaw that almost wiped out an NYC skyscraper. Slate.
- Morgenster J (1995) The fifty-nine-story crisis. The New Yorker, pp. 45–53.
- Delatte NJ (2009) Beyond failure: forensic case studies for civil engineers. ACSE Press. p. 340.

Appendix: Design-Added Value Methods Recommended for Analyzing Case Studies

Table formats useful to conduct DAV analysis. Recommended for use in conducting the analysis of case studies included in Chap. 12 through 17.

Table A.1 Building facts

Include facts like: Type, Construction phase, Design phase, Gross area, # of work spaces, Height, Levels, Cost, Gross volume, Structural system, so on; as relevant. *Indicate value for each item and the source from which you obtained that information*

Table A.2 Stakeholders

Include stakeholders like: Owner, Architects, Structural Engineers, General Contractor, Quantity Surveyor, Mechanical and Electrical, Landscape Architect, Lighting Engineer, Foundation Engineers, and so on. *Cite formal names of companies for each item*

Table A.3 Design features

Include features like: Overall Form, Shape of Footprint, Atria or Sky-gardens, Vertical spaces, Exterior cladding, Structural System, and so on. *Provide a succinct description for each item*

Table A.4 Timeline

Indicate significant dates and events for the delivery of the facility
• date-1: *government makes call on the ordinances and regulations.* • date-2: *public reaction to design* • date-3: *bids invited for the design* • … • date-*n*: *building opened to rave reviews*

Ö. Akın, *Design Added Value*, https://doi.org/10.1007/978-3-030-28860-0

Table A.5 Estimated ordinal values for stakeholder benefits

Benefits can be in categories like: *Reputation, Funds, Track record, Environmental contribution,* and so on

Table A.6 Estimated NPV for Feature; *Energy Savings* [below is sample calculation]

Floor to floor height	= 3.75 m
Surface area of facade	= (24 + 34 + 36 office wings) (3.75 m) (34 m) = 11,730 m^2
Material and construction	= (Cost per m^2) (double-skin area) = ($915.83/$m^2$) (11,730 m) = $10,742,742 in 1997$
Research investment	= $15 million in 1997 $
Operating costs	= (7 €/m^2) * ($1/(1.0919 €)) (11,730$m^2$) = $75,200 per year in 2003-$S
Cleaning costs	= (3.9 €/m^2) * ($1/1.0919 €) (11,730$m^2$) = $41,900 per year in 2003$
Total NPV costs	= [(costs in 1997) $(1.03)]^{(1997–2013)}$ +⋯+ [(costs in 2013) $(1.03)]^{(2013–2013)}$ = −$43.9 million
Electricity consumption for a conventional office building	= 15.5 kWh/ft^2
Area of floor area	= 920,314 ft^2
Ventilation	= 5% of total electricity
Space heating and cooling	= 48% of total electricity
Energy savings	= (80%) [(53%) (15.5 kWh/ft^2) (920,314 ft^2)] = $464,000 per year
NPV of energy savings	= (energy savings in 1997) $(1.03)^{(1997–2013)}$ +⋯+[(energy savings in 2013)$(1.03)]^{(2013–2013)}$ = $10.1 million

Table A.7 Estimated NPV for Feature; *Structural System* (below is sample calculation)

Size of horizontal ISMB	= (475 mm) (1100 mm) = 0.13325 m^2 cross-sectional area
Total no. of floors	= 53
Length of horizontal Vierendeel truss system at each floor	= 34 m
Number of facade sides for the Vierendeel truss system	= 3
Total horizontal length of Vierendeel system	= (53) (34) (3) = 5406 m
Total volume of horizontal member ISMB	= (cross-sectional area) ((length) = 0.13325) (5406 = 720.35 m^3
Size of vertical ISMB	= (475 mm) (1000 mm) = 0.16575 m^2 cross-sectional area
Height of each floor	= 3.75 m = 3 m without members overlapping
No. of vertical Vierendeel truss system members at each floor level	= 4 m
No. of facade sides with the Vierendeel truss system	= 3
Total vertical length of Vierendeel truss system	= (3) (4) (53) (3) = 1908 m. Total volume of vertical member ISMB = (cross-sectional area) (total length) = (0.16575) (1908) = 316.25 m^3
Total volume of steel used to construct the entire Vierendeel truss system	= 720.35 + 316.25 = 1036.6 m^3
Density of mild steel	= 7850 kg/m^3
Total density of steel used for the Vierendeel truss system	= (7850) (1036.6) = 8140 metric tons – 1 metric ton = 1.1 American tons
Total amount of steel used in Vierendeel truss system	= (8140) (1.1)= 9000 tons
Over 15 stories (including material cost + labor cost + equipment cost of ISMB per ton)	= $3500
Total cost of using Vierendeel truss system	= (1) × (2) = (9000) (3500) = $31.5 million
NPV of structural System	= (31,500,000 (1 + 0.0274) 2013–2010) = $ 34.16 million

Table A.8 Calculation of net present value of a double-skin facade over 17 years

Year	Initial investment	Operating costs	Cleaning costs	Energy savings	Annual investment	Inflation factor	Present value
1997	\$25,742,742	\$75,200	\$41,900	\$464,000	–\$25,395,842	$(1.03)^{-16}$	–\$40,752,871
1998	\$0	\$75,200	\$41,900	\$464,000	\$346,900	$(1.03)^{-15}$	\$540,459
1999	\$0	\$75,200	\$41,900	\$464,000	\$346,900	$(1.03)^{-14}$	\$524,717
2000	\$0	\$75,200	\$41,900	\$464,000	\$346,900	$(1.03)^{-13}$	\$509,434
2001	\$0	\$75,200	\$41,900	\$464,000	\$346,900	$(1.03)^{-12}$	\$494,596
2002	\$0	\$75,200	\$41,900	\$464,000	\$346,900	$(1.03)^{-11}$	\$480,191
2003	\$0	\$75,200	\$41,900	\$464,000	\$346,900	$(1.03)^{-10}$	\$466,205
2004	\$0	\$75,200	\$41,900	\$464,000	\$346,900	$(1.03)^{-9}$	\$452,626
2005	\$0	\$75,200	\$41,900	\$464,000	\$346,900	$(1.03)^{-8}$	\$439,443
2006	\$0	\$75,200	\$41,900	\$464,000	\$346,900	$(1.03)^{-7}$	\$426,643
2007	\$0	\$75,200	\$41,900	\$464,000	\$346,900	$(1.03)^{-6}$	\$414,217
2008	\$0	\$75,200	\$41,900	\$464,000	\$346,900	$(1.03)^{-5}$	\$402,152
2009	\$0	\$75,200	\$41,900	\$464,000	\$346,900	$(1.03)^{-4}$	\$390,439
2010	\$0	\$75,200	\$41,900	\$464,000	\$346,900	$(1.03)^{-3}$	\$379,067
2011	\$0	\$75,200	\$41,900	\$464,000	\$346,900	$(1.03)^{-2}$	\$368,026
2012	\$0	\$75,200	\$41,900	\$464,000	\$346,900	$(1.03)^{-1}$	\$357,307
2013	\$0	\$75,200	\$41,900	\$464,000	\$346,900	$(1.03)^{0}$	\$346,900
						NPV	–\$33,760,449

Table A.9 To conduct analyses based on the AHP ranking matrix technique applied to similar buildings. Use the format provided below to create tables *1–N* (as many as required) per each *criteria* and *feature* combination

- Use *features* like: *double-skin facade* and *structural system* and so on
- Use *criteria* like: *performance impact on occupants aesthetics* and *ease of construction* and so on

Criteria	Building-1	Building-2	Building-3	Building-4	Weighted score (WS)
Building-1	1	X	Y	Z	$1/\sum_1$
Building-2	1/X	1	1/W	1/V	$1/\sum_2$
Building-3	1/Y	W	1	U	$1/\sum_3$
Building-4	1/Z	V	1/U	1	$1/\sum_4$
$\sum$	$\sum_1 = 1 + 1/x + 1/y + 1/z$	$\sum_2 = 1 + x + y + z$	$\sum_3 = 1 + y + 1/w + 1/u$	$\sum_4 = 1 + z + 1/v + u$	$\sum_{ws} = 1/\sum_1 + 1/\sum_1 + 1/\sum_1 + 1/\sum_1 = 1.00$

Table A.10 Benchmarking of N-number of comparable buildings (example below is for four buildings) with features like: *double-skin facade, structural system,* and so on

	Building-1	Building-2	Building-3	Building-1
Location	Location-1	Location-2	Location-3	Location-4
Completion year	Year-1	Year-2	Year-3	Year-4
Height (stories)	Height-1 (stories-1)	Height-2 (stories-2)	Height-3 (stories-3)	Height-4 (stories-4)
Area	Area-1	Area-2	Area-3	Area-4
Total cost	Cost-1 in mil.$	Cost-2 in mil.$	Cost-3 in mil.$	Cost-4 in mil.$
Image	Image of building-1	Image of building-2	Image of building-3	Image of building-1

Table A.11 Ordinal analysis of impact on stakeholders

Stakeholders	1st order (construction)	2nd order (0–3 years)	3rd order (3–10 years)	*Ordinal values of*
Client	–	+	+++	−3 = three minuses −2 = two minuses −1 = one minus 0 = zero +1 = one plus +2 = two plusses +3 = three plusses
Architect	+	++	+++	
Structural engineer	+	+	++	
Facade fabricators	++	+	0	
Users	0	+++	+++	

Bibliography

AAAS (1965) The integrity of science. In: Holton G (ed) Science and culture. Beacon Press, Boston

Ackerman J (1965) On scientia. In: Holton G (ed) Science and culture, a study of cohesive and disjunctive forces. Beacon Press, Boston

Akın Ö (1978) How do architects design? In: Latombe b J (ed) Artificial intelligence and pattern recognition in computer-aided design. North Holland, New York

Akın Ö (1979) A style named post-modern. Archit Des 8-9:224–226

Akın Ö (1986/1989) Psychology of architectural design. Pion Limited, London

Akın Ö (1988) Expertise of the architect. In: Rychener MD (ed) Expert systems for engineering design. Academic, New York, pp 171–196

Akın Ö (1990a) Necessary conditions for design expertise and creativity. Des Stud 11(2):107–113

Akın Ö (1990b) Data interpretation in comprehensive design environment. In: Proceedings of the 6th international conference on systems research information and cybernetics, Baden-Baden, 12–16 August 1990

Akın Ö (1994a) Creativity in design. Perform Improv Q 7(3):9–21

Akın Ö (2004) eXtreme design. In: ARCHI

Akın Ö (2006) A Cartesian approach to design rationality. The METU Press, Ankara

Akın Ö (2008) Chapter 1: Current trends and future direction in CAD. In: Karimi H, Akıncı B (eds) CAD/GIS integration: existing and emerging solutions. Taylor and Francis, New York, NY

Akın Ö, Moustapha H (2003) Strategic use of representation in architectural massing. Des Stud 25:1

Akın Ö, Anadol Z (1993) Determining the impact of CADrafting on the building process. Int J Constr Inf Technol 2:1–8

Akın Ö, Akın C (1998) On the process of creativity in puzzles, inventions and designs. Autom Constr 7:123–138

Akın Ö, Ozkaya I (2004) Emerging CAD processes: the case of computer aided requirement management. In: Akın Ö, Krishnamurti R, Lam KP (eds) Proceedings of the GCADS-04 international symposium, 12–14 July 2004. School of Architecture, Carnegie Mellon University, Pittsburgh, PA

Akın Ö, Dave B, Pithavadian S (1992) Heuristic generation of layouts (HeGeL) based on a paradigm for problem structuring. Environ Plann B Plann Des 19:33–59

Akın Ö, Dave B, Pithavadian S (1987) A paradigm for problem structuring in design. In: Department of *Architecture Research Report Series # DA-87-01*. Carnegie Mellon University, Pittsburgh, PA

Akın Ö, Esin N, Uluoğlu B (1996) Quality of architectural service in the project management process. J Archit Plann Res 13(1):63–90

Ö. Akın, *Design Added Value*, https://doi.org/10.1007/978-3-030-28860-0

Akın Ö, Sen R, Donia M, Zhang Y (1995) SEED-pro: computer assisted architectural programming in SEED. ASCE J Archit Eng 1(4):153–161

Akın Ö, Türkaslan-Bülbül T, Garrett JH, Akıncı B, Wang H (2004) Embedded commissioning for building design. In: Proceedings of the design computing and cognition'04 conference, 19–21 July 2004. MIT Press, Cambridge, MA

Akın Ö, Türkaslan-Bülbül T, Brown S, Kim E, Akıncı B, Garrett JH, (2003) Comparison of AHSRAE guidelines with building commissioning practice. In: Proceedings of the 11th national conference on building commissioning, Palm Springs, CA, 20–22 May 2003

Akın Ö, Anadol Z, Roche J (1993) Working papers I: case studies for computer aided facilities design and management. Department of Architecture Carnegie Mellon University, Pittsburgh, PA

Akin Ö, Krishnamurti R, Lam KP (2004) Generative CAD systems. Carnegie Mellon University Press, Pittsburgh, PA

Akin Ö, Akinci B, Garrett JH Jr (2009) Identification of functional requirements and possible approaches for self-configuring intelligent building systems. In: An interim progress report for the period 07/01/2009–9/30/2009. Carnegie Mellon University, Pittsburgh, PA. Submitted to Steven Bushby, National Institute for Standards and Technology

Akin Ö, Akıncı B, Berges M, Bushby S, Garrett J, Huber D, Lee SH, Türkaslan-Bülbül T (2012) Embedded commissioning of building systems. Artech House, Inc., Boston, MA

Alexander C (1964) Notes on the synthesis of form. Harvard University Press, Cambridge, MA

Aldrich HP, Lambrechts JR (1986) Back Bay Boston, Part II: groundwater levels. Civ Eng Pract 1(2) http://www.bostongroundwater.org

Alexander C, Ishikawa S, Silverstein M, Jacobson M, Fiksdahl-King I, Angel S (1977) A pattern language: towns, buildings, construction. Oxford University Press, New York

Ali MM, Moon K (2007) structural developments in tall buildings: current trends and future prospects. Archit Sci Rev 50(3):205–223

Anderson JR (1981) Cognitive skills and their acquisition. Lawrence Erlbaum Associates, Hillsdale, NJ

Ankerl G (1981) Experimental sociology of architecture. Mouton, The Hague

Anscombe FJ, Auman RJ (1963) A definition of subjective probability. Ann Math Stat 34:199–205

Archea J (1987) Puzzle-Making: what architects do when no one is looking. In: Kalay Y (ed) The computability of design. Wiley, New York

Arend J, Benkert M, deFilippis A, Tillmaand S (2013) Commerzbank tower. In: Kheir A-K, Mir MA (eds) Tall buildings and urban design., http://www.mbenkert.com/arend.benkert.defilippis.tillmaand.pdf. Accessed Apr 2019

Aristotle (1957) The physics. Harvard University Press, Loeb Classical Library, Cambridge, MA

Aristotle (1990) Nichomachean Ethics, VI.4, 1140a. In: Cahn SM (ed) Classics of western philosophy. Hackett Publishing, Indianapolis, IN

Arlin PK (1974) Problem finding: the relation between cognitive process variables and problem-finding performance. Unpublished doctoral dissertation, University of Chicago

Arnold RA (1989) Economics. West Publishing Co., St Paul

Arup and Partners (2013/2014) Commerzbank. http://www.arup.com/Projects/Commerzbank.aspx. Accessed Apr 2019

ASSA (Architecture Steel Stahl Acier) (2010) European convention for constructional steelwork. Swiss Re, London. www.steelconstruct.com

Ayers M (1986) Locke's original atomism. In: Kenny A (ed) Rationalism, empiricism and idealism. Clarendon Press, Oxford

Bachus JW (1960) The syntax and semantics of the proposed international algebraic language for the Zurich ACM-GAMM conference. In: Information processing. UNESCO, Paris

Bacon F (1620) Author's preface. In: Novum Organum [The new organon, or true directions concerning the interpretation of nature]

Baer A (1976) A cognitive model of problem formulation in design. In: Institute of physical plant reports. Carnegie Mellon University, Pittsburgh, PA

Baglivo JA, Graver JE (1983) Incidence and symmetry in architecture and design. Cambridge University Press, Cambridge

Ballard DH, Brown CM (1982) Computer vision. Prentice-Hall, Englewood Cliffs

Ballard G (2008) The lean project delivery system: an update. Lean Constr J:1–19. http://www.leanconstruction.org/lcj/paper_2008_issue.html

Barker RG (1968) Ecological psychology. Stanford University Press, Stanford, CA

Bashkar R (1982) Emergence, explanation and emancipation. In: Secord PF (ed) Explaining human behavior. Sage Press, Beverly Hills, CA

Battles SJ, Burns EM (2000) Trends in building-related energy and carbon emissions: actual and alternate scenarios. In: Summer study on energy efficiency in buildings conference, American Council for an Energy-Efficient Economy, 21 August 2000

Bauman JF (2000) From Tenements to the Taylor Homes: in search of an urban housing policy in twentieth-century America. Pennsylvania State University Press, University Park, PA

Baumol WJ (1972) Economic theory and operations analysis. Prentice-Hall, Englewood Cliffs, NJ

Baykan CA, Fox MS (1991) Constraint satisfaction techniques for spatial planning. In: ten Hagen PJW, Veerkamp PJ (eds) Intelligent CAD systems III: practical experience and evaluation. Springer-Verlag, Berlin, pp 187–204

Beck K, Boehm B (2003) Agility through discipline: a debate. IEEE Comput 36(6):44–46

Beck K (2004) eXtreme programming explained. Addison-Wesley, Boston

Beck P (2001) The AEC dilemma: exploring the barriers to change. In: Proceedings of design intelligence. Greenway Communications and Design Futures Council, Washington, DC

Beedle LC (1988) Second century of the skyscraper. Van Nostrand Reinhold Company, New York

Bjork B (1991) Intelligent Front-ends and product models. Artif Intell Eng 6(1):47–55

Bligh Office Tower (2011) ingenhoven architects 00:00. https://www.archdaily.com/169173/1-bligh-office-tower-ingenhoven-architects. Accessed Apr 2019

Bon R (1989) Building as an economic process. Englewood Cliffs, Prentice Hall

Brady S (2015) Citicorp Center Tower: how failure was averted. Eng J

Bridwell SG (2010) Swiss Re by Norman Foster. In: Studio 489, 200. http://web.utk.edu/~archinfo/a489_f02/Phase1.html

Broadbent G (1973) Design in architecture. In: Architecture and the human sciences. Wiley, Chichester

Broadbent G, Ward A (1969) Design methods in architecture. In: Architectural Association Paper Number 4. George Wittenborn Inc, New York

Brown FE, Steadman JP (1987) The analysis and interpretation of small house plans: some contemporary examples. Environ Plann B 14:407–438

Bruner JS, Goodnow JL, Austin GA (1956) A study of thinking. Wiley, New York

Buchanan P (2007) The tower: an anachronism awaiting rebirth. Harv Des Mag. www.gsd.harvard.edu/research/publications/hdm/back/26_Buchanan.pdf. Accessed Nov 2010

Byrne O (ed) (1847) The first six books of the elements of euclid. Chiswick Press, London

Calhoun JB (1962) A behavioral sink. In: Bliss E (ed) Roots of behavior. Paul Hober, New York

Campbell R (1995) Builder faced bigger crisis than falling windows. The Boston Globe, March 3

Caroll JM, Thomas JC, Malhotra A (1978) Presentation and representation in design problem solving. Br J Psychol 71:143–153

Wharton C, Rieman J, Lewis C, Polson P (1994) The cognitive walkthrough: a practitioner's guide. In: Nielsen J, Mack RL (eds) Usability inspection methods. Wiley, New York

CEIR (2019) This rising star of Perth's skyline is setting a new building sustainability standard. In: Create engineering ideas into reality https://wwwcreatedigitalorgau/perths-skyline-setting-building-sustainability-standard/. Accessed Apr 2019

Chadwick GF (1961) Works of Sir Joseph Paxton. Architectural Press, New York

Chan CS (1990) Psychology of style in design, unpublished PhD Dissertation, Department of Architecture, Carnegie Mellon University, Pittsburgh, PA

Chase WG, Simon HA (1973) The mind's eye in chess. In: Chase WG (ed) Visual information processing. Academic Press, New York, pp 215–282

CII (2009) Applicability of CII best practices by industry sector and project type. In: Construction Industry Institute, 3925 West Braker Lane (R4500), Austin, TX. https://www.construction-institute.org/scriptcontent/r-teams-list.cfm?section=res

Clark R, Pause M, Twenty Students of the School of Design (1979) Analysis of precedent: an investigation of elements, relationships, and ordering ideas in the work of eight architects, The student publication of the School of Design, North Carolina State University at Raleigh

Cockburn A (2002) Agile software development: software through people. Addison-Wesley, Harlow

Collins P (1965) Chapter 19: Rationalism. In: Changing ideals in modern architecture. McGill-Queen's University Press, Montreal, pp 198–217

Coombs CH (1964) A theory of data. Wiley, New York

Coyne R (1991) ABLOOS: an evolving hierarchical design framework. Unpublished PhD Dissertation, Department of Architecture, Carnegie Mellon University, Pittsburgh, PA

Coyne RD, Gero JS (1985) Design knowledge and sequential plans. Environ Plann B 12:419–442

Cross N (1984) Developments in design methodology. Wiley, New York

Crouch EAC, Wilson R (1982) Risk/benefit analysis. Ballinger Publishing Co., Cambridge MA

Cuff D (1992) Architecture: the story of practice. The MIT Press, Cambridge, MA

Cushman RF (1985) Property management handbook. Wiley, New York

Dalley J (1982) Design creativity and the understanding of objects. Des Stud 3(3):133–137

Darke J (1979) The primary generator and the design process. Des Stud 1(1):36–44

Davis (1995a) Prototyping. Adv Comput 40:39–62

Davis AM (1995b) Software prototyping. In: Yovits MC (ed) Advances in computers, vol 40. Academic Press, New York, NY, pp 39–42

Debajyoti P, Cheol-Soo P, Godfried A (2006) Roles of building performance assessment in stakeholder dialogue in AEC. Autom Constr 15(4):415–427

deGroot AD (1965) Thought and choice in chess. Mouton, The Hague

deGroot AD (1966) Perception and memory versus thought: some old ideas and recent findings. In: Kleinmuntz J (ed) Problem solving. Wiley, New York, pp 19–50

Delatte NJ (2009) Beyond failure: forensic case studies for civil engineers. ACSE Press, Virginia, pp 333–345

DLBLS (2010) Department of Labor, Bureau of Labor Statistics, Consumer Expenditures in 2008, Report 1023, March 2010, Tables A and B, p 2

Drexler A (1977) The Architecture of the Ecoles des Beaux Arts, October 29, 1975–January 4, 1976. The Museum of Modern Art

Durand J-N-L (1980) Précis des Leçons d'Architecture. Ecole Polytechnique, Paris

Earl CF (1980) Rectangular shapes. Environ Plann B Plann Des 7:311–342

Eastman C (1999) Building product models: computer environments supporting design and construction. CRC Press, New York

Eastman C, Teicholz P, Sacks R, Liston K (2008) Handbook: a guide to building information modeling for owners, managers, designers, engineers and contractors. Wiley, Hoboken, NJ

Eastman C (1968) Explorations of the cognitive processes in design. Research Report, Department of Computer Science, Carnegie-Mellon University

Eastman C (1969) Cognitive processes and ill-defined problems: a case study from design. In: Proceedings of the first joint international conference on artificial intelligence, Washington, DC, pp 675–699

Eastman C (1970) On the analysis of intuitive design processes. In: Moore GT (ed) Emerging methods in environmental design and planning. The MIT Press, Cambridge, MA, pp 21–37

Eastman C (1973) Automated space planning. Artif Intell 4:41–64

Eckersley M (1988) The form of design process: a protocol analysis study. Des Stud 9(2):86–84

EIA (2003) 2003 EIA Commercial Building Energy Consumption Survey (CBECS). http://www.eia.gov.

EIA (2008) Energy Information Administration, Annual Energy Review 2008, Table 2.5, p 52

Einstein A, Infeld L (1938) The evolution of Physics. Simon and Schuster, New York

Ekman P (1973) Darwin and facial expression: a century of research in review. Academic, New York

Elbibany H, Bechtel J, Brach B, Ault D (1997) Facility management value-adding functional analysis model. ASCE J Archit Eng 3. https://doi.org/10.1061/(ASCE)1076-0431(1997)3:4(170)

Emery FE (1969) Systems thinking. Penguin Books, Ltd, Harmondsworth, Middlesex

Ericsson KA, Simon HA (1984) Protocol analysis: verbal reports as data. The MIT Press, Cambridge, MA

Fann KT (1970) Peirce's theory of abduction. Nijhoff, The Hague

Fedun W (2005) 30 St. Mary Axe by Foster and Partners

Feldman J, Lindell MK (1990) On rationality. In: Horowitz I (ed) Organization and decision theory. Kluwer Academic Publishers, Boston, pp 83–165

Feyrabend PK (1878) science in a free society. New Left Books, London

Fischer V (1997) Sir Norman Foster and Partners Commerzbank, Frankfurt am Main. Axel Menges, Stuttgart

Fisk WJ (2005) Health-related costs of indoor ETS, dampness, and mold in the United States and in California. In: Proceedings of indoor air conference 2005, Beijing, 4–9 Sep 2005, pp 308–313

Fisk WJ (2000) Health and productivity gains from better indoor environments and their relationship with building energy efficiency. Annu Rev Energy Environ 25:537–566

Fitch JM (1966) American building, the historical forces that shaped it. The Riverside Press, Cambridge, MA. Second edition revised and enlarged (1947, first edition)

Fjeld T, Veiersted B, Sandvik L, Riise G, Levy F (1998) The effect of indoor foliage plants on health an discomfort symptoms among office workers. Indoor Built Environ 7:204–209

Flemming U (1986) On the representation and generation of loosely-packed arrangements of rectangles. Environment and Planning B: Planning and Design 14:323–350

Flemming U (1988) Rule-based systems in computer-aided architectural design. In: Rychener M (ed) Expert systems for engineering design. Academic Press, Boston, pp 93–112

Flemming U, Chien SF (1995) Schematic layout design in the SEED environment. J Archit Eng 1(4):162–169

Flemming U, Bhavnani SK, John BE (1997) Mismatched metaphor: user vs. system model in computer-aided drafting. Des Stud 18:349–368

Foster and Partners (2013) Commerzbank Headquarters. Architravel; Online Architecture Guide http://www.fosterandpartners.com/projects/commerzbank-headquarters/ Accessed Apr 2019

Foster and Partners (2012) Hearst Tower. In: Flashback: Archdaily, http://www.archdaily.com/204701/flashback-hearst-tower-foster-and-partners/ . Accessed Apr 2019. Awards by Project 2010, http://www.fosterandpartners.com/Data/Awards/S.aspx. Accessed Oct 2010

Foz A (1973) Observation on the designer behavior in the parti. In: The design activity international conference, volume I, printing unit. University of Strathclyde, Glasgow, pp 19.1–19.4

Frampton K (1987) Modern architecture, a critical history. Thames and Hudson Ltd, London. Second printing of the revised and enlarged edition

Freeman PA, Newell A (1971) A model for functional reasoning in design. In: Proceedings of the second international joint computer conference on artificial intelligence. British Computer Society, London, pp 621–640

Friedman AT (2006) Women and the making of the modern house. Yale University Press, New Haven, CT, p 130

Fuller S (2003) Kuhn vs Popper. Icon Books Ltd, London

Galle P (1995) Towards integrated "intelligent" and compliant computer modeling of buildings. Autom Constr 4(3):189–211

Gamma E, Helm R, Johnson R, Vlissidis J (1995) Design patterns: elements of reusable object-oriented software. Addison-Wesley, New York

Garrett JH, Kılıççöte H, Choi B (1995) Providing formal support for standards usage within SEED. J Archit Eng 1(4):187–194

Gartner (2012) Commerzbank Tower. http://josef-gartner.permasteelisagroup.com/project-gallery/475

Gerard RW (1946) The biological basis of imagination; with biological sketches. Sci Mon 62:477–499

Getzels JW (1982) Without questions, what would the answer be? In: Hogarth RM (ed) Question framing and response consistency. Jossey-Bass Inc Publishers, San Francisco, pp 37–50

Ghirardo D (1988) Desperately seeking decentering. Archit Rev J 183(1096):4–6

Goncalves J, Bode K (2011) The importance of real life data to support environmental claims for tall buildings. In: Council on tall buildings and urban habitat journal, Issue II

Goodman N (1972) Problems and projects. Bobbs-Merill, Indianapolis, IN

Goodman N, Elgin K (1988) Reconceptions in philosophy and other arts and sciences. Hackett, Indianapolis, IN

Goumain PG (1973) Design methods and designer's methods. In: The design activity international conference, volume 1, printing unit. University of Strathclyde, Glasgow, pp 23.1–23.8

Grason J (1970) A dual linear graph representation for space filling location problems of the floor plan type. In: Moore GT (ed) Emerging methods of environmental design and planning. The MIT Press, Cambridge, MA, pp 172–181

Grattan-Guinness I (1970) The development of foundations of mathematical analysis from Euler to Riemann. The MIT Press, Cambridge, MA

Grayling AC (1995) Philosophy 1: a guide through the subject. University of London, London

Gregory S (1966) The design method. Butterworth Inc, London

Gropius W (1965) The new architecture and the Bauhaus (paperback edition). The MIT Press, Cambridge, MA

Guilford JP (1950) Creativity. Am Psychol 5:444–454

Gutman R (1988) Architectural practice, a critical view. Princeton Architectural Press, Princeton, NJ

Haldane ES, Ross GT (1911) The philosophical works of Descartes, 2 volumes. Cambridge University Press, Cambridge

Hall ET (1966) The hidden dimension, A doubleday anchor book. Garden City, NY

Handelma S (1983) Jacques Derrida and the heretic hermaneutic. In: Krupnick M (ed) Displacement: Derrida and after. University of Indiana Press, Bloomington, IN

Hartkopf V (2013) Towards a sustainable future: integrating environmentally and economically future proof technologies and practices. In: Proceedings of the NSF/IUCRC annual meeting, Washington, DC

Hawking S (1988) A brief history of time, from the big bang to black holes. Bantam Books, New York

Hayes JR (1982) Issues in protocol analysis. In: Decision-making: an interdisciplinary inquiry. Kent Publishing Company, Boston, MA, pp 61–63

Heath T (1984) Method in architecture. Wiley, New York

Heath T (1956) Euclid. In: The thirteen books of euclid's elements. Dover, New York

Heerwagen J, Zagreus L, Zagreus JH (2005) The human factors of sustainable building design: post occupancy evaluation of the Philip Merrill environmental center. In: Indoor environmental quality (IEQ) series. Center for the Built Environment, Center for Environmental Design Research, UC Berkeley

Hensen JLM (1995) On system simulation for building performance evaluation. In: Proceedings of the 4th IBPSA world congress on building simulation '95, Madison, WI, pp 259–267

Hillier B, Hanson J (1984) The social logic of space. Cambridge University Press, Cambridge

Hitchcock HR, Johnson P (1966) The international style. W W Norton and Company, Inc, New York

Hobhouse C (1950) 1851 and the crystal palace. John Murray, London

Hoffman A (n.d.) Why they built the Pruitt–Igoe project. In: Joint center for housing studies. Harvard University, Cambridge, MA

Hogarth RM (1982) Question framing and response consistency. Jossey-Bass Inc, San Francisco, CA

Holmes E (1878) The life of Mozart including his correspondence. Chapman and Hall, London, p 211
Hong T, Chou SK, Bong TY (2000) Building simulation: an overview of developments and information sources. Build Environ 35(4):347–361
Horowitz I (1990) Organization and decision theory. Kluwer Academic, Boston, MA
Howard R (2006) CAD, curved surfaces and building quality. J Inf Technol Constr 11:427–436
Hunt EB (1975) Artificial intelligence. Academic Press, New York
Hutchins EL (1995) Cognition in the wild. The MIT Press, Cambridge, MA
IAI (2003) Early design IFC. Construction engineering research laboratory, http://www.iai-na.org/technical/early_design.php. Accessed 15 Dec 2003
ITECH (2011) Interior design support. http://web.ku.edu/~itech/index.html. Accessed 21 Apr 2011
Jackendoff R (1983) Semantics and cognition. The MIT Press, Cambridge, MA
Jencks C (1977) Language of post-modern architecture. Rizzoli International Publications, Inc, New York
Jones JC, Thornley DC (1963) Conference on design methods. Pergamon Press, Oxford
Jones JC (1970) Design methods: seeds of human futures. Wiley, New York
Kahneman D, Tversky A (1979) Prospect theory: an analysis of decision under risk. Econometrica 47:263–291
Kamara JM, Anumba CJ, Evbuomwan NFO (2000) Computer-based application for the processing of clients' requirements in construction. J Comput Civ Eng 14(4):264–271
Kant I (1958) Critique of pure reason, translation and introduction by Norman Kemp Smith, University of Edinburgh, Abridged Edition. The Modern Library, New York
Kenny A (1986) Rationalism empiricism and idealism. Oxford Press, Oxford
Kerlinger FB (1979) Behavioral research: a conceptual approach. Holt, Rinehart and Winston, New York
Kifner J (1973) Boston Tower's plywood windows. The New York Times, ARCHIVES, 18
Knight TW (1990) Mughal gardens revisited. Environ Plann B Plann Des 17:73–84
Krauss RI, Myer RM (1970) Design: a case history. In: Moore GT (ed) Emerging Methods in Environmental Design and Planning. The MIT Press, Cambridge, MA, pp 11–20
Kretschmer E (1931) The psychology of men of genius. Routledge and Kegan Paul Ltd, London
Krishnamurti R (1980) The arithmetic of shapes. Environ Plann B Plann Des B 7:463–484
Krishnamurti R, Giraud C (1986) Towards a shape editor: the implementation of a shape generation system. Environ Plann B 13(4):391–404
Kuhn T (1962) The structure of scientific revolutions, International Encyclopedia of Unified Science, Vol. 2, Number 2, 2nd edn. University of Chicago Press, Chicago
Kviniemi A, Fischer M (2003) Requirements management with product models. In: Proceedings of the nD workshop. Salford University, Manchester
Lambot I (1997) Commerzbank Frankfurt: prototype for an ecological high rise. Watermark Birkhäuser, Boston, MA
Landau R (1997) Architecture, ethics, and the person. In: Pollak M (ed) The education of the architect. MIT Press, Cambridge, MA
Langdon D (2005) Spon's architects' and builders' price book, 2005, 130th edn. Spon Press, New York
Lawrence HO, Samuel HW (2009) Purchasing power of money in the United States from 1774 to 2010. In: Measuring worth
LBNL (2011a) Impacts of building ventilation on health and performance. In: IAQ-SFRB indoor air quality scientific findings resource bank, Lawrence Berkeley National Laboratory. Available at: http://www.iaqscience.lbl.gov/vent-summary.html
LBNL (2011b) Indoor dampness, biological contaminants and health. In: IAQ-SFRB indoor air quality scientific findings resource bank, Lawrence Berkeley National Laboratory. Available at: http://www.iaqscience.lbl.gov/vent-summary.html. Accessed 21 Apr 2011
Corbusier L (1929) The defense of architecture. Stavba VII:10

Corbusier L (1986) Towards a new architecture, (translated from Vers Une Architecture by Frederick Etchelles). Dover Publications, Inc., New York

Doll R, Peto R (1976) Mortality in relation to smoking: 20 years' observations on male British doctors. Br Med J 2:1525–1536.

Leaman A (2001) Productivity improvement. Buildings in value, building use studies, Vol. 3, Ch. 19

Lee SH, Akin O (2010) Augmented reality-based computational support for operations and maintenance fieldwork. Autom Constr

Lee W (1971) Decision theory and human behavior. Wiley, New York

Leffingwell D, Widrig D (2000) Managing software requirements: a unified approach. Addison-Wesley, Reading, MA

Lehman A (2005) Within these walls. Safety and Health Practitioner. http://www.shponline.co.uk/. Accessed 31 October 2019

Lemay L (2004) Prescriptive specification versus performance specifications. Concrete Monthly. http://www.concretemonthly.com/monthly/art.php?884

LePatner BB (2008) It's time to fix America's broken construction industry. Engineering News-Record, McGraw-Hill Construction

Lesnikowski WG (1982) Rationalism and romanticism in architecture. McGraw-Hill, Inc., New York

Levy DNL (1990) How will chess programs beat Kasparov? In: Marsland TA, Schaeffer J (eds) Computers chess and cognition. Springer-Verlag, New York

Levey M, Salvadori M, Woest K (1994) Why buildings fall down: how structures fail. W. W. Norton & Company

Lewis RK (1985) Architect? In: A candid guide to the profession. The MIT Press, Cambridge, MA

Liggett R (1992) A designer-automated algorithm partnership, an interactive graphic approach to facility layout. In: Kalay YE (ed) Evaluating and predicting design performance. John Wiley and Sons, pp 101–124

Lindsay PH, Norman DA (1972) Human information processing. Academic Press, New York

Liu G, Liu M (2011a) Comparison of measured cooling and heating consumption with initial simulated values and initial cooling and heating calibration signatures. Build Environ 46(2):409–420

Liu G, Liu M (2011b) A rapid calibration procedure and case study for simplified simulation models of commonly used HVAC systems. Build Environ 46:409–420

Locke J (1986) An essay concerning human understanding. In: Nidditch PH (ed) Oxford, 1975, XV, 5 in "Locke and the ethics of belief". In: Kenny A (ed) Rationalism, empiricism and idealism Clarendon Press, Oxford

Lynch K (1960) The image of the city. The Technology Press And The Harvard University Press, Cambridge, MA

Mackey J (2010) Evaluating work spaces in European sustainable high rise buildings-assessments, responses and speculations from the inhabitants. University of Florida

Mackinder M, Marvin H (1982) Design decision-making in architectural practice. Research Paper 19, Institute of Advanced Architectural Studies, University of York, The King's Manor, York

Mackinnon DW (1970) The personality correlates of creativity: a study of American architects. In: Vernon PE (ed) Creativity. Penguin Books Ltd., Harmondsworth, Middlesex, pp 289–311

Maher ML (1988) HI-RISE: an expert system for preliminary structural design. In: Rychener M (ed) Expert systems for engineering design. Academic Press, Boston, MA, pp 37–52

Manicas PT (1987) A history and philosophy of the social sciences. Basil Blackwell, Oxford

March L (1976) The architecture of form. Cambridge University Press, Cambridge

March L, George S (1985) Spatial systems in architecture and design: some history and logic. Environ Plann B 12(1):31–53

Marshall RD, Pfrang EO, Leyendecker EV, Woodward KA, Reed RP, Kasen MB, Shives TR (1982) Investigation of the Kansas City Hyatt Regency Walkways Collapse. (NBS BSS 143), Building Science Series – #143, May 31, [Washington DC.]. https://digital.library.unt.edu/ark:/67531/metadc38324/m1/35/. Accessed December 28, 2020. University of North Texas Libraries,

UNT Digital Library, https://digital.library.unt.edu; crediting UNT Libraries Government Documents Department

Matteo J, Nash K, Kin M (1996) Frank Lloyd Wright's Fallingwater: a non-destructive structural investigation & analysis. In: Proceedings for the third annual NDE conference, Boulder, Colorado

McCain I (2009) DiaGrid: structural efficiency and increasing popularity. Presentation. http://www.dsg.fgg.uni-lj.si/dubaj2009/index.php?limitstart=20. Accessed November 2010

McDermott J (1982a) R1: a rule-based configurer of computer systems. Artif Intell 19:39–88

McDermott J (1982b) Domain knowledge and the design process. Des Stud 3(1):31–36

McKean J (1994) Crystal palace: Joseph Paxton & Charles Fox. Phaidon Press, London

Meehan E (1975) Public housing policy. Center for Urban Policy Research, Rutgers-The State University, New Brunswick, pp 64–65

Mendel AO (1947) The living thoughts of Descartes presented by Paul Valery. David McKay Company, Philadelphia, PA

Messent D (1997) Opera house act one. D. Messent Photography, Unit 17, 28 Rosenberry Street, Balgowlab, NSW 2093, Sydney Australia

Millar CJ (1986) Frank Lloyd Wright letters to clients the press. California State University, Fresno, CA, p 82

Miller GA (1956) The magical number seven, plus or minus two. Psychol Rev 63:2

Miller GA, Galanter E, Pribram KH (1960) Plans and the structure of behavior. Henry Holt, New York

Mitchell W (1977) Computer-aided architectural design. Van Nostrand Reinhold, New York

Mitchell WJ (1979) Synthesis with style. In: Proceedings of the international conference on the application of computers in architecture, building design and urban planning, Berlin, May, pp. 120–134

Mitchell WJ (1990) The logic of architecture. The MIT Press, Cambridge, MA

Mitchell WJ, Steadman P, Liggett RS (1976) Synthesis and optimization of small rectangular floor plans. Environ Plann B Plann Des 3(1):37–70

Mogulescu M (1970) Profit through design. American Management Association, Inc, New York

Montgomery R (1985) Pruitt–Igoe: policy failure or societal symptom. In: The metropolitan midwest urbana. University of Illinois Press, Chicago, IL

Moon K, Connor JJ, Fernandez JE (2007) Diagrid structural systems for tall buildings: characteristics and methodology for preliminary design. Struct Design Tall Spec Build 16(2):205–230

Moore GT (1970) Emerging methods in environmental design and planning. The MIT Press, Cambridge, MA

Moore H (1955) Notes on sculpture. In: Ghiselin B (ed) The creative process. Mentor Books, New York

Morgenstern J (1995) The fifty-nine-story crisis. The New Yorker, May 29, pp. 45–53

Morrin N (2008) Sustainable landmark for Swiss Re in London. http://skanska-sustainability-case-studies.com/index.php/Sustainable-Landmark-for-Swiss-Re-in-London.html. Accessed April 2019

Mudge A (1989) Value engineering. J. Pohl Associates, New York, NY

Munro D 2004 Swiss Re's building, London. Nyheter om Stalbyggnad.NR3. 36–43. www.epab.bme.hu/microstation/FreeForm/Examples/SwissRe.pdf. Accessed November 27, 2010

Nabelek A, Nabelek I (1985) Room acoustics and speech perception. In: Katz J (ed) Handbook of clinical audiology. Williams and Wilkins, Baltimore, MD, pp 834–846

Nagel E, Newman JR (1958) Gödel's proof. Routledge and Kegan Paul Ltd., London

Newell A (1968) On the analysis of human problem solving protocols. In: Gardin JC, Jaulin B (eds) Calcul et Formalisation dans les Sciences de L'Homme. Centre National de la Recherche Scientifique, Paris

Newell A (1970) Heuristic programming: ill-structured problems. In: Arnofsky JA (ed) Progress in operations research, vol 3. John Wiley, New York, pp 360–414

Newell A, Simon HA (1972) Human problem solving. Prentice-Hall, Englewood Cliffs, NJ

Newell A, Shaw JC, Simon HA (1958) Elements of a theory of human problem solving. Psychol Rev 65:3

Newman O (1972) Defensible space. Collier Books, New York

Newmarch R (1906) Life and letters of Peter Ilich Tchaikovsky, John Lane, pp 274–275

Nielsen J (1994) Usability engineering. Morgan Kaufmann, San Francisco

Nozick R (1974) Anarchy, State and Utopia. Basic Books, New York

O'Connor J (2004) Survey on actual service lives for North American buildings. In: Woodframe housing durability and disaster issues conference, Las Vegas, October

O'Hear A (1989) Introduction to the philosophy of science. Clarendon Press, Oxford

Oldroyd D (1986) The arc of knowledge. Methuen, New York

Page J (2000) My crown of thorns. The Guardian: London. http://www.guardian.co.uk/uk/2000/may/04/dome.millennium. Accessed April 2019

Pan X, Han CS, Dauber K, Law KH (2006) Human and social behavior in computational modeling and analysis of egress. Autom Constr 15(4):448–461

Pelli C (2019) Pelli Architects. Torre de Cristal 2019. http://pcparch.com/project/torre-de-cristal/. Accessed April 2019

Pennick N (1979) The ancient science of geomancy, living in harmony with the earth. CRCS Publications, Sebastopol, CA and Thames and Hudson Ltd., London

Peterson CR, Beach LR (1967) Man as an intuitive statistician. Psychol Bull 68:29–46

Petroski H (1982) To engineer is human. St. Martin's Press, New York

Pevsner N (1949) Pioneers of modern design, New York

Pfrang EO, Marshall R (1982) Collapse of the Kansas City Hyatt regency walkways. Civil Eng ASCE 52(7):65–69

Piaget J (1947) The psychology of intelligence, translated from French by Malcolm Piercy. Rautledge and Kegan Paul, London

Piggott J (2004) Palace of the people, London

Poincaré H (1962) Science and method by F Maitland (tr). Dover, New York, p 51

Poirazis H (2004) Double skin facades for office buildings. Department of Energy and Building Design, pp 77–78

Polanyi M (1995) Personal knowledge: toward a post-critical philosophy. Harper Torchbooks, New York

Polya G (1973) How to solve it, a new aspect of mathematical method. Princeton University Press, Princeton, NJ

Popper KR (1963) Conjectures and refutations: the growth of scientific knowledge. Routledge and Kegan Paul, London

Popper KR (1972) Objective knowledge. Oxford University Press, Oxford

Powell K (2006) 30 St Mary axe: a tower for London. Merrell Publishers, London/New York

Pratt JW, Raiffa H, Schlaifer R (1964) The foundations of decision under uncertainty: an elementary exposition. J Am Statist Assoc 59:353–343

Preiser WFE, Rabinowitz HZ, White ET (1998) Post occupancy evaluation. Van Nostrand Reinhold, New York

Preiser, Wolfgang, Hardy, Andrea E., Schramm, Ulrich (Eds.) (2018) Building Performance Evaluation From Delivery Process to Life Cycle Phases. Springer International Publishing. DOI 10.1007/978-3-319-56862-1

Schramm, Ulrich (Eds.) (2018) Building Performance Evaluation From Delivery Process to Life Cycle Phases. Springer International Publishing. DOI 10.1007/978-3-319-56862-1

Proshansky HM, Ittelson WH, Rivlin LG (1970) Environmental psychology: man and his physical setting. Holt, Rinehart and Winston, New York

Radford A, Gero J (1986) Design by optimization in architecture and building. Van Nostrand Reinhold, New York

Rahman A, Mahmood AK (2008) Agent-based simulation using prometheus methodology in evacuation planning. In Proceedings from international symposium on information technology, Kuala Lumpur, August 26–28, Vol. 3, pp. 1–8

Randall JH (1962) The career of philosophy. Columbia University Press, New York
Reinhardt J (2010) Computer-aided construction management. November 30, 2010, Guest speaker at course 12–611. Carnegie Mellon University, Pittsburgh, PA
Reitman WR (1964) Heuristic decision procedures, open constraints and structure of ill-defined problems. In: Shelly MW, Bryan GL (eds) Human judgments and optimality. John Wiley, New York, pp 282–315
Revesz GE (1988) Lambda-calculus, combinatorics and functional programming. Cambridge University Press, Cambridge, MA
Richards W (1988) Natural computation. The MIT Press, Cambridge, MA
Rittel H, Webber M (1973) Dilemmas in a general theory of planning. Policy Sci 4:155–169
Rogers CR (1970) Towards a theory of creativity. In: Vernon PE (ed) Creativity. Penguin Books Ltd., Harmondsworth, Middlesex, pp 137–152
Rossbach S (1983) Feng-Shui—the Chinese art of placement. E. P. Dutton, New York
Rossiter J (2006) The gherkin up for sale for £600m. This is Money. co.uk. http://www.thisismoney.co.uk/markets/article.html?in_article_id=412713&in_page_id=3
Rowe P (1987) Design thinking. The MIT Press, Cambridge, MA
Ruchelman LI (1977) The world trade center. Syracuse University Press, Syracuse
Rush R (1986) The building systems integration handbook. John Wiley and Sons, New York
Ruskin J (1981) The seven lamps of architecture, Farrar, Straus and Giroux, New York
Russell B (1996) The principles of mathematics, 2nd edn. W W Norton and Company, New York. (first published in 1902)
Russo JE, Dosher BA (1976) An information processing analysis of binary choice. Carnegie-Mellon University, Pittsburgh, PA
Saaty T (1980) The analytic hierarchy process: planning, priority setting, resource allocation. McGraw-Hill, New York, NY
Saint A (1983) The image of the architect. Yale University Press, New Haven
Sarin RK (1990) Analytical issues in decision methodology. In: Horowitz I (ed) Organization and decision theory. Kluwer Academic Publishers, Boston, MA, pp 13–63
Savage LJ (1954) The foundations of statistics. John Wiley and Sons, New York
Schoemaker PJH (1980) Experiments on decisions under risk: the expected utility hypothesis. Kluwer-Nijhoff Publishing, Boston, MA
Schön DA (1983) The reflective practitioner: how professionals think in action. Basic Books, New York
Schwartz TA (2001) Glass and metal Curtain-Wall fundamentals. APT Bull 32(1):37–45
Scruton R (1980) The aesthetics of architecture. Princeton University Press, Princeton, NJ
Scruton R (1981) From Descartes to Witgenstein, a short history of modern philosophy. Routledge and Kegan Paul, London
Scruton R (1983) The aesthetic understanding. Methuen, London
Seppänen O, Fisk WJ, Lei QH (2006) Effect of temperature on task performance in office environment. Report, Lawrence Berkeley National Laboratory, Environmental Energy Technologies Division, Indoor Environment Department, Berkeley, CA
Shen Y (2009) Green Design and the City, Buildings: The Gherkin, London. October 15. www.greendesignetc.net/Buildings_09/Building_Shen_Yuming_paper.pdf Accessed November 2010
Shenkel WM (1977) Real estate finance American institute of banking. American Bankers Association, Washington, DC
Simon H (1973) The structure of ill-structured problems. Artif Intell 4:181–201
Simon HA (1957) Models of man: social and rational. Wiley, New York
Simon HA (1970) Style in design. In: Proceedings of the environmental design research association conference II. Department of Architecture, Carnegie-Mellon University, Pittsburgh, PA, pp 1–10
Simon HA (1969/1981) The sciences of the artificial. The MIT Press, Cambridge, MA

Simon HA, Feigenbaum EA (1964) An information-processing theory of some effects of similarity, familiarization, and meaningfulness in verbal learning. J Verbal Learn Verbal Behav 3:385–396

Skinner S (1982) The Living Earth Manual of Feng-Shui. Chinese Geomancy Routledge and Kegan Paul, London

Slovic P, Fischhoff B, Lichtenstein S (1977) Behavioral decision theory. Annu Rev Psychol 28:1–39

Smith A (1976) In: Todd WB (ed) An inquiry into the nature and causes of the wealth of nations (first published in 1776) RH Campbell and AS Skinner (general eds), vol 2. Clarendon Press, Oxford

Smith A (1980) In: Wightman WPP, Byrce JC (eds) Essays on philosophical subjects. DP Raphael and AS Skinner (general eds). Clarendon Press, Oxford

Sommer R (1969) Personal space: the behavioral basis of design. Prentice Hall, New York

Sommer R (1974) Tight spaces; hard architecture and how to humanize it. Prentice-Hall Inc., Englewood Cliffs, NJ

Stanford Encyclopedia of Philosophy (2003) Bayes' Theorem. http://plato.stanford.edu/entries/bayes-theorem/. First published Sat Jun 28, 2003; substantive revision Tue Sep 30, 2003

Stauffer LA, Ullman DC (1988) A comparison of the results of empirical studies into the mechanical design process. Des Stud 9(2):107–114

Steadman JP (1983) Architectural morphology. Pion Ltd., London

Stevens G (1990) The reasoning architect, mathematics and science in design. McGraw-Hill Publishing Co., New York

Stillings NA, Feinstein MA, Garfield JL, Rissland EL, Rosenbaum DA, Weisler SE, Baker-Ward L (1987) Cognitive science: an introduction. The MIT Press, Cambridge, MA

Stiny G (1976) Two exercises in formal composition. Environ Plann B Plann Des 3:187–210

Stiny G (1985) Computing with form and meaning in architecture. J Archit Educ 39(/1):7–19

Stiny G (1990) What is design? Environ Plann B Plann Des 17:97–103

Stiny G, Gips J (1972) Shape grammars and the generative specification of painting and sculpture. Inf Process 71:1460–1465

Stone PA (1983) Building economy design production and organization: a synoptic view, 3rd edn. Pergamon Press, Oxford

Stübe K, Utzon J (2009) Sydney opera house: a tribute to Jørn Utzon. Reveal Books

Suh NP (2001) Axiomatic design, advances and applications. Oxford University Press, Oxford

Sullivan L (1896) Kindergarten chats and other essays

Sullivan L (1956) The autobiography of an Idea (republication of the original by the American Institute of Architects, 1924). Dover Publications, Inc., New York

Teicholz P (2004) Labor productivity declines in the construction industry: causes and remedies. AECbytes Analysis, Research, Reviews, Web Publication, Issue #4 (April 14) http://www.aecbytes.com/viewpoints.html

Thieme A, Schröter F (2013) Modern heavy steel plates for use in offshore installations – characteristics and production processes. Steel Construction Wiley Online Library. https://onlinelibrary.wiley.com/doi/abs/10.1002/stco.201310025. Accessed April 2019

Toker F (2003) Fallingwater rising. Alfred A Knopf, New York

Turbayne CM (1962) The myth of metaphor. Yale University Press, New Haven and London

Türkaslan-Bülbül T, Akın Ö (2006) Computational support for building evaluation: embedded commissioning model. J Autom Constr 15(4):438–447

Tversky A (1969) Intransitivity of preferences. Psychol Rev 76:31–48

Tversky A (1972) Elimination by aspects: a theory of choice. Psychol Rev 79:281–299

Tversky A, Kahneman D (1974) Judgments under uncertainty: heuristics and biases. Science 185:1124–1131

USDT-FHA (2004) Performance specifications strategic roadmap: a vision for the future. United States Department of Transportation - Federal Highway Administration Publication, Spring 2004

Valery P (1947) The living thoughts of *Descartes*. David McKay Company, Philadelphia, PA

Venturi R (1966) Complexity and contradiction in architecture, the museum of modern art papers on architecture. The Museum of Modern Art, New York

Vernon PE (1970) Introduction. In: Vernon PE (ed) Creativity. Penguin Books Ltd., Harmondsworth, Middlesex, pp 9–16

Viollet-le-Duc EE (1959) Discourses on architecture. Grove Press, Inc., New York

Viollet-le-Duc EE (1987) Lectures on architecture. Dover Publications Inc, New York

Vitruvius (1960) The ten books on architecture (republication of the Morris Hicky Morgan translation; originally published by Harvard University Press, 1914). Dover Publication, Inc., New York

von Neumann J, Morgenstern O (1953) Theory of games and economic behavior. Princeton University Press, Princeton, NJ; 1944, second ed. 1947, third ed. 1953

Waldron MB, Waldron KJ (1989) A time sequence study of a complex mechanical system design. Des Stud 9(2):95–106

Warszawski A, Rosenfeld Y (1994) Robot for interior-finishing works in building: feasibility analysis. J Constr Eng Manag 120:1

Watson A (ed) (2013) Building a masterpiece—the Sydney opera house – lessons in space and environment. Powerhouse Publishing, Sydney

Watson PC, Johnson- Laird PN (1972) Psychology of reasoning, structure and content. Harvard University Press, Cambridge, MA

Webb EJ (1966) Unobtrusive methods. Rand McNally, Chicago, IL

Werner J (2014) The design flaw that almost wiped out an NYC skyscraper. Slate

Westre A (2008) Complexity machine 1: a 3D modeling application implementing behavioral simulation. In: Proceedings of the ACADIA 2008: Silicon + Skin: biological processes and computation conference, October 16–19

Whitehead B, Eldars MZ (1964) The planning of single story layouts. Archit J 139:1373–1380

Whole Life Costing 2004 Table-X—importance of maintenance and operational costs (based on 2002–3 typical cost figures by BMI). Published by Consulting Excellence. www.constructing-excellence.org.uk. Accessed November 2010

Wikipedia the Free Encyclopedia (2011) Sensitivity analysis. http://en.wikipedia.org/wiki/Sensitivity_analysis. Modified on 25 May 2011 at 15:10

Williams B (1995) Ethics. In: Grayling AC (ed) Philosophy. Oxford University Press, Oxford, pp 545–582

Williamson SH (2009) Seven ways to compute the relative value of a U.S. dollar amount, 1774 to present. Measuring Worth. http://www.measuringworth.com/calculators/inflation/result.php Accessed April 2010

Wilson R (1990) Foreword. In: Ruegg RT, Marshall HE (eds) Building economics: theory and practice. Van Nostrand Reinhold, New York, NY

Woodbury RF, Chang TW (1995) Massing and enclosure design with SEED-Config. J Archit Eng 1(4):170–178

Woolley K (2010) Reviewing the performance the design of the Sydney opera house. The Watermark Press, Boorowa, NSW

Wyatt MD, et al (1853) Exhibition of works of all nations in 1851. Collection of printed documents and forms. Vol 2. Royal Commission for the Exhibition of 1851, London National Art Library Pressmark: Ex.1851.93

Zhao F, Mary LM (1988) Using analogical reasoning to design buildings. Report EDRC-12-22-88. Engineering Design Research Center, Carnegie-Mellon University, Pittsburgh, PA

Zimring C, Mahbub R, Kampschroer K (2010) Facility Performance Evaluation (FPE). Public Buildings Service. U.S. General Services Administration, last updated: June 11, 2010

Zsambok CE, Klein G (1997) Naturalistic decision-making. Lawrence Erlbaum Associates, Mahwah, NJ

Index

Ö. Akın, *Design Added Value*, https://doi.org/10.1007/978-3-030-28860-0

W

Z